MOTHS OF THE PAST

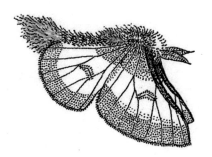

"IT'S ALL PART OF THE SAME THING."

— J. S. RABKIN (with apologies to Parmenides and Heraclitus)

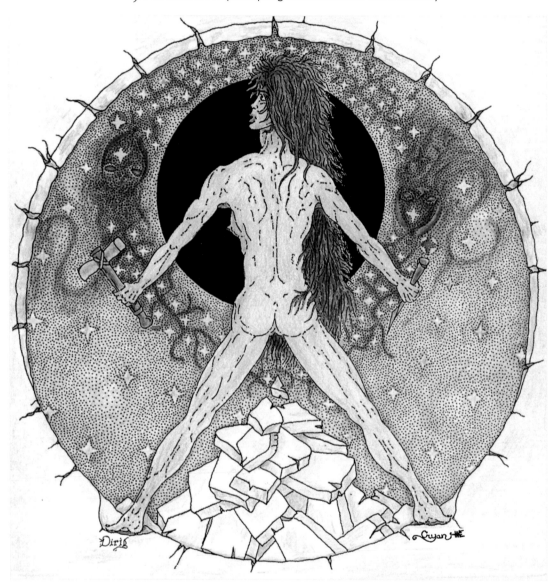

"A WORLD TURNED ON ITSELF"

▲ FRONTISPIECE: *Escaping Plato's Cave*: In this nod to LEONARDO DA VINCI's *Vitruvian Man*, the young Eidolon *Maia* (Mother) confronts the Monster *Chaos* with its Duality *Paradox*, whose unleashed forces, *Time* and *Space*, create and devour one another. [Also see pp. 15 & 39.]

◄ FRONT COVER: *Two Natives Meet in the Primordial Fen*: On a sunny autumn morning, 5000 years ago, a young proto-Iroquoian woman discovers a dramatic new moth in her territory, while gathering stems of Tawny Cottongrass. She also notices an egg ring on Sweet Gale, and a dead sprig of Tamarack. The feathers in her twisted deerskin headband are of Broad-winged Hawk, Wild Turkey, and Common Crow, the three totemic birds of her people. Her bracelet is made from dried seed capsules of Buckbean.

ഇ൚ൠ

MOTHS OF THE PAST

EASTERN NORTH AMERICAN BUCK MOTHS (*HEMILEUCA*, SATURNIIDAE), WITH NOTES ON THEIR ORIGIN, EVOLUTION, AND BIOGEOGRAPHY

John F. Cryan & Robert Dirig

ℰℭ

CONTENTS

*Two Natives Meet
in the Primordial Fen* ◆ Front Cover

Frontispiece: *Escaping Plato's Cave*
◆ facing page 1

Front Matter ◆ 1
Abstract ◆ 1
Prologue ◆ 2

Part One

Introduction ◆ 3-4
Methods ◆ 4-5
Results and Interpretation ◆ 5-7

Part Two

Paradigm Shift ◆ 8-9
The "Theory of Everything" ◆ 9-14
Prelude to Convolution ◆ 15-18
Building Better Buck Moths:
The Other Half of Evolution ◆ 18-22
Reconstructing Linnaeus ◆ 23-25

Part Three

Hemileuca iroquois (*H. maia* ×
H. nevadensis), *Semispecies
nova prima* ◆ 26-32
Conclusion: A Plea to Humanity ◆ 33-36
Ten More Commandments for Humanity
and Our Only Home ◆ 36
Epilogue: New Atlantis, 2258 ◆ 37-38

Addenda ◆ 39-42
Authors & Illustrators ◆ 43
Afterword & Acknowledgements ◆ 43-44

Tailpiece: *To the Edge of the World*
◆ facing page 44

June Nocturne in the Fen ◆ Back Cover

ℰℭ

ABSTRACT: The BOG BUCK MOTH, *Hemileuca iroquois* (*H. maia* × *H. nevadensis*) Cryan & Dirig, *S. n. p.*, **2020** (Saturniidae), is described as new. This rare, diurnal **Semispecies** (introduced here as a new taxonomic category) occurs in a few relict, alkaline fens on the eastern edge of Lake Ontario in New York, USA, and Ontario, Canada. It selects BUCKBEAN (*Menyanthes trifoliata*, Menyanthaceae), an emergent aquatic herb, as its larval **Starter Host**. The new moth originated about 8000 years ago from natural crossing of proto-*H. maia* and proto-*H. nevadensis* in a postglacial genetic **Fluxus** that was centered (and still expands) in the Great Lakes Basin of the Upper Midwest. From this locus, *H. iroquois* moved eastward through calcareous wetlands associated with the bed of **Glacial Lake Iroquois**, reaching its present location about 5000 years ago.

The complicated, three-year process of understanding the **Great Lakes Buck Moth Fluxus** involved moving beyond **nodal thinking** to examine **continua** and **chaos**, incidentally producing unexpected insights that extend the landmark historical contributions of Linnaeus, Darwin, and Einstein. The physical agent of chaos is the minuscule **Chaon**, which fills the three-dimensional matrix of expanding and accelerating spacetime. Everything in the Universe is made of *chaons*, which are the agents of most variation upon which natural selection acts. Darwin's "survival of the fittest" is the *destructive* half of evolution, while **Convolution** is its key *creative* force — the cause of evolutionary progress in complexity, diversity, and overall fitness of organisms. It also resolves the **Species Problem**. Contemplation of paradoxes and mirroring was a major theme of this process. Cryan's **Chaon-Convolution Theory** is introduced and explained.

[KEY WORDS are in **boldface**.]

Don Rittner, Publisher
*Pine Bush Historic Preservation Project
Occasional Publication No. 2.*
ISBN 978-0-578-66012-7
[Please see recommended citation on page 41.]
The authors may be reached at:
Moths of the Past, P. O. Box 225, Hensonville, NY 12439-0225, USA

ℰℭ

PROLOGUE

THIS PUBLICATION FEATURES A HIGHLY DIVERSE COMBINATION OF CONCEPTS AND PRESENTATION STYLES that thoroughly blend art with science. *It contains much that is new.* Ostensibly foremost is our original description of a remarkable new saturniid moth, which is characterized in *Parts One* and *Three*. Perhaps of greater importance is a contextual re-sorting, synthesis, and extension of long-held wisdom from many areas of human endeavor, including mythology, mathematics, biology, physics, chemistry, astronomy, history, religion, biogeography, geology, genetics, taxonomy, cosmic and organic evolution, and conservation of our Earth and its amazing (and now imperiled) envelope of life. These are elucidated in *Part Two*, the *Conclusion*, and the *Epilogue*. Taken together, this is *a highly philosophical work*. We understand that many of these innovative ideas may surprise, and possibly delight, puzzle, or even alarm some readers.

Because this discussion is very intricate and intense, some sections may be difficult to grasp on a first reading. We have endeavored to present this information in an accessible fashion by supplementing the prose with diagrams, maps, photographs, cartoons, decorations, and cover art; and by text references to page numbers of illustrations and similar subjects. Rereading *Part Two* and the *Conclusion* several times will help integrate the new ideas. The *Epilogue* presents a futuristic glimpse of Earth's eventual rescue and recovery.

Re-examining scientific aspects through a wide-angled artistic lens has helped conceptualize these thoughts. The text was conceived and drafted by John, who offers these interpretations as springboards for further research and philosophical development. In addition to participating in the early field work, discovery, and habitat documentation of our moth, Bob has contributed intellectual discussion, editorial services, layout and design, and art. We remain astounded and humbled by the wealth of insights and corollaries that have resulted from the process of describing a single new moth.

ಸಾಡಾ

Buckbean's three-parted leaves (×0.5) emerge from a matrix of Large Cranberry. The *inset* shows Buckbean's intricate, white-fringed flowers (×1.2).

Part One

INTRODUCTION

IN AUTUMN **1977**, the same year that DON RITTNER's Pine Bush Historic Preservation Project published *Moths of Autumn* (our attempt to save the Buck Moths at the vanishing Albany Pine Bush — p. 19), we found a mysterious Buck Moth specimen in the Cornell University Insect Collection, and tracked its origin to the eastern shore of Lake Ontario. It looked like a perfect mashup of the then-three described species — the classic, narrow-banded BUCK MOTH of the East (*Hemileuca maia*), the translucent NORTHERN BUCK MOTH of New England (*Hemileuca lucina*), and the wide-banded WESTERN BUCK MOTH of western North America (*Hemileuca nevadensis*). The following year, we revisited the remnant, postglacial floating fens

(pp. 27, 29, & 41) that were sequestered behind lakeshore dunes, where we had found the moths flying and spotted their egg rings, nine months before. The summer fens were filled with relict northern vegetation. We immediately spied clumps of black Buck Moth caterpillars feeding on a most unusual host, BUCKBEAN (*Menyanthes trifoliata*; p. 2 & back cover), a monotypic, circumboreal plant of the Buckbean family (Menyanthaceae), with succulent leaves that resembled giant clovers emerging from fleshy, pale rhizomes running beneath the waterlogged fen mats. Being wholly herbaceous above water, some of the plants had died down for the winter before the moths flew in autumn, but were in full leaf and flower in June.

We spent the following few years exploring the bed of former GLACIAL LAKE IROQUOIS (p. 5) and its tributaries and outlets (the largest region of mostly alkaline soils in New York State), looking for additional fens spared by agriculturalization and urbanization that might harbor the strange Buck Moth. Although some interesting, intact bogs and fens were found, including a few with Buckbean, the moths were absent. During this time, we learned that Canadian naturalists had discovered two populations of *Menyanthes*-feeding Buck Moths in southern Ontario fens. After they heard of our discoveries, we were graciously hosted by ROSS A. LAYBERRY, and J. DONALD LAFONTAINE of the Canadian Museum of Nature in Ottawa, while visiting these sites, which were far larger and in more pristine condition than the ones in adjacent N.Y.

As the eastern shore of Lake Ontario was under strong developmental pressure at this time, we notified CAROL RESCHKE of the New York Natural Heritage Program (a partnership between New York's Department of Environmental Conservation and The Nature Conservancy, a non-profit land preservation organization) of our discovery. At this point we had found an additional moth population in a fen about 20 miles inland from the Lake. Further searching by us and others had revealed no other large colonies of the moth in New York, but a few smaller ones were disclosed near the originally located Lake sites.

Over the next four decades, due in part to the impact of *Moths of Autumn* and a follow-up article by Cryan in *Defenders* magazine ("Retreat in the Barrens," Jan.-Feb. 1985, pp. 18-29), which detailed the plight of *H. maia* in the disappearing Pitch Pine Barrens of the Northeast, many individuals and organizations around the U.S. began studying this fascinating group. What emerged in aggregate from this work were conservation efforts for many of the remaining Northeastern populations, and others elsewhere in the U.S., including major populations of the mystery moth in N.Y. and Ontario, and some of its Great Lakes relations; descriptions of new Buck Moth species in the greater Texas region; and a growing focus on perplexing taxonomic issues involving an expanding area around the Great Lakes and Upper Midwest. ℘∝

METHODS

E HAD BOTH BEEN AVID SATURNIID REARERS as teenagers, and Cryan had already initiated crossing studies of Buck Moth populations from the Northeast and elsewhere, earlier in the 1970s. Prior to that, he had reared and crossed dozens of species of native and exotic giant silkmoths, during a golden age of amateur insect collecting, trading, and rearing that flourished in the early post-WW-II era. He learned that it was possible to cross many congeneric species, especially closely related "sibling species." The hybrids were mostly sterile, but occasionally they were fertile enough to produce a generation or two. And very rarely, certain intergeneric crosses could produce offspring too.

The most spectacular and longest-lasting result was obtained by crossing what was called the GLOVER'S SILK MOTH (*Hyalophora gloveri*), sent by DUKE DOWNEY of Sheridan, Wyoming, with the CECROPIA MOTH (*H. cecropia*) from Long Island, N.Y. The resulting "red-banded maroon" hybrids were beautiful, fantasy moths of that group, intermediate in size between the parent species, and fertile enough to be carried forward on their own, without backcrossing, for quite a few generations.

Another insight that emerged was that the **founder effect** was variable, at least in captive breeding and crossing work. The best example was "the-moths-that-would-not-die," a culture of the EMPEROR MOTH (*Saturnia pavonia*) from Ireland, kept casually for over a decade. There was a lot of inbreeding, some years between only one pair. But the population always immediately bounced back into the hundreds.

After re-reading the writings of DARWIN and MENDEL, particularly passages on animal husbandry and plant hybridization, and later works bringing domestication and agricultural history and practices up to date, Cryan began a trial-and-error effort to factory-farm *Hemileuca*.

It settled on exclusive use of indoor rearing inside metal larval containers, borrowed from the work of Cornell Entomology Professor JOHN G. FRANCLEMONT. Stock was kept in large, inner-city building basements, usually in close proximity to heating, where temperatures of 75º-80ºF. and low humidity could be maintained year-round. To accommodate the larger and more active Buck Moth caterpillars, two sizes of old metal herbarium boxes (11¾ ×16½ inches, and 3 or 5¼ inches high) were substituted for the smaller, circular movie reel containers used by Franclemont to raise noctuid larvae. Their lids were propped open by an inner lining of fine-mesh wire screening. Upon hatching, larvae were given a salad of leaves, usually including oak (*Quercus*), willow (*Salix*), cherry (*Prunus*), Purple Loosestrife (*Lythrum salicaria*), *Spiraea*, and *Menyanthes*. They were fed their chosen host, but often weaned off Loosestrife and *Menyanthes* onto one of the others, as early as possible after the second molt. Larval density was thinned as they broke cluster, usually after the third and subsequent molts. Pupation was in wooden boxes with screened lids over milled peat moss (p. 43). Adult development and emergence were timed and triggered using manipulated light, temperature, and moisture cues.

Attempts were made to mate pairs naturally under artificial light, using fans to move sex pheromones, but results were inconsistent. Hand pairings without further manipulation, other than the sometime presence of a calling female from the same population, were also spotty. Knock-out hand pairing was finally tried, using low-dose (older) cyanide jars, and this proved satisfactory for most population combinations, though it was tricky. Eggs were stored over the winter in old (pre-self-defrosting) refrigerators to maintain adequate humidity. Initial stock was usually obtained as wild-collected egg rings. Continuous base stocks of Northeastern populations, and intermittent ones for several years of other populations, were kept and bred. As many replicates as possible were made of each cross.

The idea was to create a standardized artificial moth-breeding environment in which the loss of instinct through domestication would include the stripping of many behavioral responses to variable natural cues, thus neutralizing most pre-zygotic reproductive isolating mechanisms. In addition, year-to-year seasonal scheduling conformity would tend to synchronize metamorphic events (egg hatching, larval feeding period, pupation period, and adult emergence), allowing moths from different latitudes to be crossed.

By the early 1980s, full-scale, slow-motion production (one generation per year) had commenced. It ran through 2011-2012, when back-to-back Hurricanes Irene and Sandy destroyed much of the equipment and stock.

ഇഈ

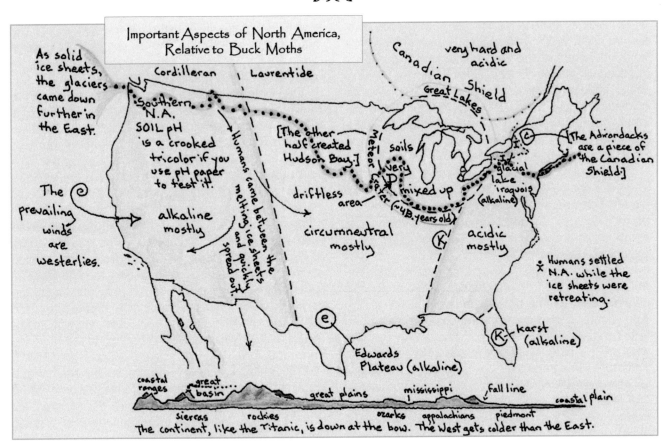

Important Aspects of North America, Relative to Buck Moths

As solid ice sheets, the glaciers came down further in the East.

Cordilleran | Laurentide

Southern N.A.

SOIL pH is a crooked tricolor if you use pH paper to test it.

The prevailing winds are westerlies.

Canadian Shield
very hard and acidic

Great Lakes

[The other half created Hudson Bay.]

Humans came between the melting ice sheets and quickly spread out.

driftless area

Meteor crater

soils very mixed up

(~4B. years old)

glacial lake iroquois (alkaline)

[The Adirondacks are a piece of the Canadian Shield.]

alkaline mostly

circumneutral mostly

acidic mostly

Humans settled N.A. while the ice sheets were retreating.

Edwards Plateau (alkaline)

karst (alkaline)

coastal ranges | great basin | great plains | mississippi | fall line | coastal plain

sierras | rockies | ozarks | appalachians | piedmont

The continent, like the Titanic, is down at the bow. The West gets colder than the East.

RESULTS AND INTERPRETATION

ENOUGH CROSSES WERE SUCCESSFUL to create maps showing *ley lines* of Buck Moth historical *consanguinity*, and *shadow vectors* of their postglacial movements, generally northward out of Ice Age refugia (p. 7). The three older named entities and the new moth largely maintained species-like internal cohesion among crossed populations. But results from the large region circumscribing the Great Lakes were the confusing exception, with shadow vectors indicating past radial movements in this area.

There seem to have been *two main postglacial migratory thrusts of the* primordial (proto-*nevadensis*) *Western Buck Moth* — one arcing through riparian habitats relatively quickly, north-ward out of Mexico, southern California, and Nevada, west of the high Rockies, occupying the Great Basin; then crossing the high mountains where LEWIS and CLARK did, many thousands of years later, in the opposite direction, reaching southern Canada, and then curving down the Prairie Potholes and terminating in the *Wisconsin Drift-less Area*, which had escaped glaciation during the last Ice Age. The other moved north more slowly, following riverbeds through the drier eastward rain shadow of the Rockies, and petered out in Wyoming. These lines may have originated from one large refugium with *maia*-interactive satellites.

The ancestors of the *nominate (proto-*maia*) Buck Moth of the East* definitely had two refugia. One was centered in the greater Texas-Mississippi Delta region and its surrounding exposed cont-inental shelf, which, at that time, was narrow on the northern edges of the Gulf of Mexico. The second included southern Florida, the Bahamas, and a wide, exposed region of the Atlantic continental shelf that joined them. The two refugia may have been connected near the beginning of the Wis-consinan glacial maximum, but tenuously at best. They were separate by the start of glacial retreat.

Four major migratory pathways of drought- and fire-adapted *Eastern Buck Moths* (proto-*maia*) moved north as the glaciers melted back: three out of the eastern Texas-Louisiana-Mississippi Delta refugium, and one from the Florida-Bahamas. They moved at varying speeds, the fastest being west of the Mississippi, and the slowest laboriously hill-hopping up the Appalachians. The northward column east of the Mississippi, but west of the Appalachians, also encountered obstacles in terrain and dense, wetter forests, compared with the one west of the river, which traversed drier, less hilly, and more open habitats.

The *Atlantians* had co-evolved away from main-line Buck Moths, when they seized upon a small-leaved, shrubby oak with high tolerance for acids, salts, metals, and cold, which had invaded and colonized the deep, coarse, recently exposed sands of the oceanic continental shelf and adjacent coastal plains just inland. They rode variations of that starter host all the way up the expanded east coast, through bleak, rapidly colonizing, and de-pauperate heaths, moors, and dwarf pine plains that were ravaged by fires, and punctuated by enormous dunes, blowouts, and rivers running to the sea. This oak's small acorns were carried north, against periglacial winds, much faster than any other oak's, by jays, crows, and Passenger Pigeons.

Just before the onset of the postglacial period of (natural) maximum warming known as the *Xero-thermic* or *Hypisthermal Interval*, about 8000 years ago, *Hemileuca* moths from the east, coming north, met others from the west, coming south, in the Wis-consin Driftless Area. A *hybrid zone* formed, even-tually producing varyingly fertile entities that com-bined the whole genomes of proto-*maia* and proto-*nevadensis*. An expanding *fluxus* ensued, like the ripples in a pond, in which some of these entities continuously crossed and backcrossed with each other and with the parental entities, which before initial re-contact, had been separated for at least 100,000, perhaps over 200,000, generations (*i.e.*, years). That fluxus, though broken up by agriculture and development in many places, persists to this day, and continues to expand in area (pp. 7, 26, 35). Some of the most easily observed hallmarks of it across populations are the widely varying band widths and opacity of the moths (p. 31), great variations in caterpillar livery, and the adoption of novel starter hosts and new habitats.

One of those entities began feeding on the abun-dant *Menyanthes*, which had recolonized rapidly out of its Driftless Area refugium, and from the south, into almost every freshwater body left behind in glaciated terrain. In the early postglacial era, cold water cov-ered more than half of the landscape in the wake of glacial retreat, slowly diminishing as the millenia advanced and the climate ameliorated. A flexible *base-swarmer* (pp. 28, 41), also with the well-devel-oped ability of Buck Moths to use diverse *finishing hosts* in later instars, if it defoliated its chosen *starter host*, the new moth began moving east — breaking free of the fluxus, and rapidly populating the newly forming alkaline wetlands created by the bulldozing actions, deposition, and rushing melt-waters of the wasting and receding Laurentide Ice Sheet (pp. 7, 26, 35).

This entity would eventually crash against the highly acidic eastern Canadian Shield, some 3,000 years later, ending its expansion (p. 7). *Menyanthes* is pH-bimodal, but is far more luxuriant in alkaline fens than in acidic bogs. Also, compared with *sphag-num bogs*, which have cold water all year, sedge- and rush-dominated *fens* are *heat sinks* that concentrate warmth in summer and early fall.

What was this mystery moth, and what caused the *fluxus*, whose disordered progeny litter the Midwestern landscape to this day, expanding and diversifying still, even into the face of extinction?

෴

Ley Lines of Historical Consanguinity

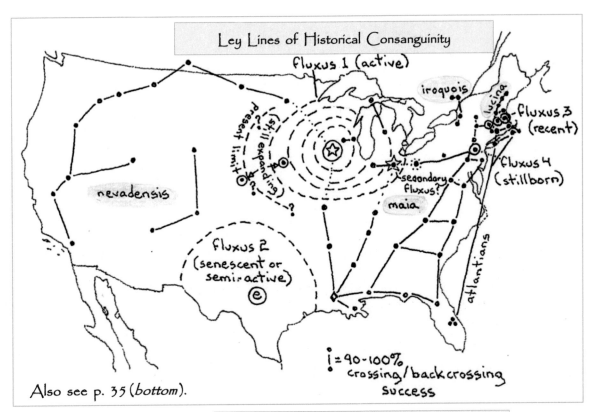

fluxus 1 (active)

iroquois

lucina

fluxus 3 (recent)

present limit?

still expanding?

fluxus 4 (stillborn)

nevadensis

secondary fluxus?

maia

fluxus 2 (senescent or semi-active)

atlantians

│ = 90-100% crossing/backcrossing success

Also see p. 35 (*bottom*).

Shadow Vectors of Postglacial Movements

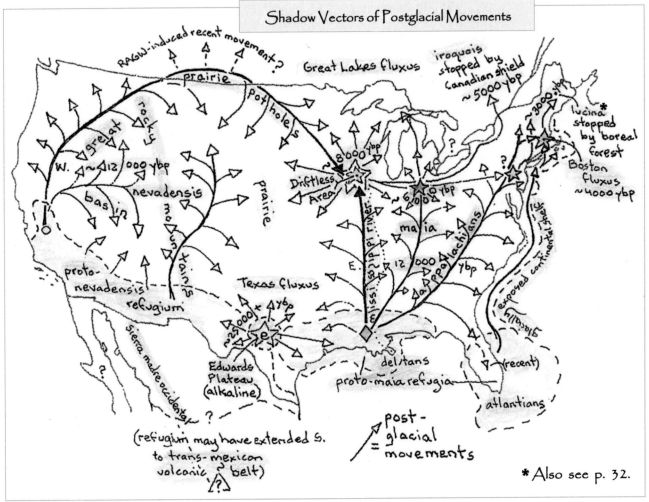

RAGW-induced recent movement?

prairie

Great Lakes fluxus

iroquois stopped by Canadian shield ~5000 ybp

3000 ybp

lucina stopped by boreal forest

great rocky

potholes

W. ~Δ12,000 ybp

prairie

8000 ybp

Driftless Area

Boston fluxus ~4000 ybp

basin

nevadensis

mountains

~6000 ybp

maia

proto-nevadensis refugium

Texas fluxus

E. Mississippi River

Appalachians

~12,000 ybp

exposed continental shelf

glacial

Sierra madre ocidental?

~2,000± ybp

Edwards Plateau (alkaline)

deltans

proto-maia refugia

(recent)

atlantians

(refugium may have extended S. to trans-mexican volcanic belt)

▷ post-glacial movements

* Also see p. 32.

7

Part Two

PARADIGM SHIFT

FTER A PERIOD OF REFLECTION on the Buck Moth fluxus question, which was recognized as a manifestation of the "Species Problem" writ small, Cryan decided the answer lay in casting a wider net. So began the laborious process of *gap analysis*, followed by an attempt to fill some of the biggest ones. This took three years.

The first part was to *examine Human limitations, and those of science*, following this logical progression:

♦ *The greatest Human capacity is for love*, or affinity, in all its forms. *Our greatest fear is death and the chaos that causes it*, and in life, *irrelevance or loneliness*. And our *greatest limitation is nodal, or typological thinking*. All three of these are deeply hard-wired in.

♦ *All Human activity, expression, and a-chievement is art*. While we celebrate the works of amazing individuals crowned with the term "renaissance," almost everything we do and create is done together. The most important reason is that art cannot exist without a maker and a user. This is the basis of economy, and ecology, and more fundamentally, society. Community.

♦ And, *for all Earthly life, symbiosis is the fundamental process* (p. 9). The gentle, flexible and compatible molecules of life all have *symbiotic hooks*. So does anything made of them, at any scale or degree of complexity.

♦ *All of life is nodal*, made of these molecules, which in turn are made of smaller parts. But all life processes, and all processes in general, are continuous, governed by the master continuum of existence: *spacetime*.

♦ The *master duality of existence*, therefore, is the *continuum-node paradox*. And the *master process* is the *paradox of attraction and repulsion*.

♦ Being largely *visual and auditory animals*, our *two primary Human modes of artistic expression are language and mathematics. Our languages, broadly speaking, are visual and auditory*. They include words, numbers, music, visual and performing arts, crafts, tools (now called *technology*), clothing (costume and adornment), shelter, and all Human-made objects, and the stories and other forms made from these — especially fantasies or invented ones — that don't seem to exist in those forms in observable Nature. All of this is collectively called *culture*. Our cultures constantly shift, evolve, and interact in time and space, following a complex interplay of events, natural and human actors, and new technology remembered as history. *Agriculture* is a set of arts vital to humanity. *Warfare* and its antecedent, *hunting*, aren't, anymore. *Cooking and eating* are in their own category, along with just *socializing*. So is our *sexuality*, which has developed far beyond reproduction.

♦ *Science*, at its core, is *a set of unitary and social arts and processes aimed at arriving at truth* (however partial or tentative) about nature, including our own, through observation, conception, testing, measuring and validation, rejection, or modification by repetition.

♦ *Mathematics*, at its base, is an *abstract art*. It has utility, in that numbers and symbols of numbers (*variables*) can tag real nodal clusters, and even points on continua, from emptiness to infinity. But *most of it lives in itself, and has no relation to object reality*, instead *creating its own, abstract version*. And *its equations, by their very nature, have a disorienting, mirror-like quality*.

♦ *Words, numbers, symbols, and objects are nodal*. They appear solidly in the moment, where they solidify our nodal bias.

♦ ALBERT EINSTEIN conveyed his concept of the *master continuum*, *spacetime*, and the *matter-energy equivalency* before that, using the symbolic power of mathematics. But he conceived *relativity* using visualization of *continua* of related activities. *By finding a nearly perfect mathematical solution for the regular workings of the Universe at scale, he stunted science*. Forevermore, scientists would pursue only order, using mathematics. *This enshrined our nodal bias as crowning achievement*.

♦ Thus *the biggest gap in science is our ignorance of chaos. The language of science (mathematics) cannot capture it*. Randomness, yes; true patternless disorder, no. It also *cannot accurately represent the actions of more than a handful of continua at once*, especially in shifting relations to one another. Perhaps most revealing is that the numerical approaches best suited to mirroring the devilish details of reality in any depth turn out to be the most daunting and exotic forms — the mathematics of approximation, estimation, probability, restriction, and transformation. These limitations are reflected in the gaps in scientific knowledge, and paradoxically *amplify our ignorance of chaos, some of the biggest and scariest examples of which arise from order that is overly complex*.

◆ Can one do **science without mathematics** (many a school-kid's dream at some point)? The answer is *yes*, if one cheats by allowing counting and measuring, which led to geometry, our first step out of the cave. **Counting and measuring are the most practical applications of mathematics**.

The only solution to this conundrum is to **escape the prison of nodal thinking and denial, based on fear of death**. This is harder than it sounds, for our brains are nodal, made of nerve cells and synapses. And so are our computers, made of electronic versions of these. That won't change. What's needed is **scientific imagination**, a form of **disciplined daydreaming**. And the **answers found need to be expressed in words (and pictures)**, because words *elide*; numbers don't. Good word combinations expand in the mind like popcorn does in the stomach. Those qualities make *words* better than our more exact, and therefore nodal *numbers*, at expressing continuity. **Mathematical precision and seeming robustness can often mislead**, especially at the beginning of exploring a question, or in attempts to study constantly changing multi-continuum processes that are too complex for math to model. **Data block** and **false rigor** are more and more often leading us down blind alleys. Computers are accelerating this trend.

◆ The trick was to invent and practice **mock-continuum thinking**. This was a bit like iterative meditation, or contemplation with an objective. The ultimate Aristotelian exercise. One had to go down the path blazed by EINSTEIN, as he looked out his window at the clock, turm, and trains from his patent office in Bern.

◆ Once started, it became clear, using this method, that **reality can best be represented as a megadimensional matrix of infinite numbers and types of continua interacting in myriad and ever-shifting ways**, which can never be fully understood by nodal beings and their nodal machines. It also became clear that to progress past our scientific gaps at all, we need a paradigm shift, a new form of science: a science of multiple and varying outcomes that fully acknowledges and incorporates continua, and relies in part on inferences that themselves may only be partially and obliquely testable.

◆ What emerged after three years of following this sort of thought were two linked theories: the **Chaon-Convolution Theory**. ౩౦౧౩

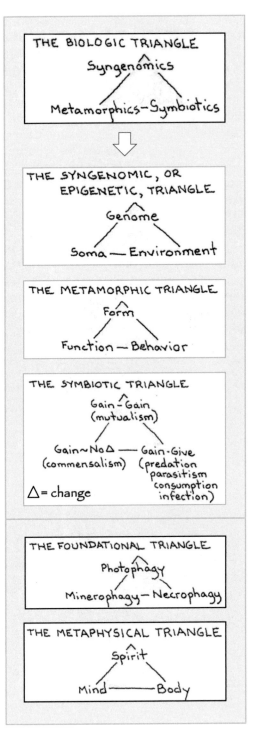

THE "THEORY OF EVERYTHING"

T STARTED WITH THE RECOLLECTION OF DARWIN'S FAMOUS UNANSWERED QUESTION in his Notebook B: "**Why is life short**" [with no question mark]. That, and the **causes of variation**, remained elusive to him.

The pathway opened by the rediscovery of MENDEL's work supposedly answered the second question, but not the first. A century and a half of additional work has left that question standing, as well as its paradoxical companion: **What causes evolutionary progression**; *i.e.*, the increase in complexity and capability of organisms over time? A subquestion of this is **why did evolutionary progress happen on an exponential time scale**; *i.e.*, accelerate?

The **biggest problems in cosmic physics** loomed at the same time. **Where's the gravity particle? What's all that dark matter and energy made of? What goes on inside a black hole? And why, with so much orderly explanation, is space still so disorderly?**

There must be **a physical agent of the continuum of chaos and its effects**: discomfort, disorder, degradation, decay, and death (and dust!). And so there is. It's called the **Chaon**. And in the ultimate paradox, **it is the fundamental particle of nature**. The **agent of most variation upon which natural selection acts**. The God particle. And the Lucifer particle, too. Everything is made of chaons. That includes us.

Chaons (*below*) **are the much-sought Gravitons**. They form the three-dimensional matrix of expanding and accelerating spacetime. And they dance to EINSTEIN's laws, while mischievously winking at SCHROEDINGER's infamous collapsing equation and other quantum weirdness.

Why haven't chaons been found? Because they're so tinyprimal that they're behind a functional, structural, and scalar barrier. They can't be detected by anything we can build out of them and their many-layered products. If such a machine existed, it would wreck spacetime, and therefore itself. So chaons can only be inferred by their effects across all scales.

When the atom smashers first started up, what came out was a veritable particle zoo. Most of the detected bits and pieces were highly unstable and very short-lived. Eventually scientists honed in on a few of them, and created a sort of "periodic chart" of subatomic particles. They called these fundamental, but they're not. They're only the first few layers. Their machines are too weak to break them apart any further, and if they do, their detection apparatus cannot find the sub-sub-subparticles, because they are already behind the scalar part of the barrier as well; *i.e.*, they're too small and/or short-lived to be observed with the equipment we have.

The Chaonic Mandala

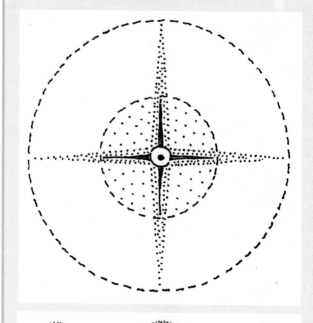

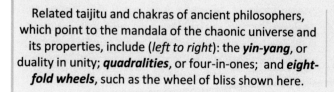

Related taijitu and chakras of ancient philosophers, which point to the mandala of the chaonic universe and its properties, include (*left to right*): the **yin-yang**, or duality in unity; **quadralities**, or four-in-ones; and **eight-fold wheels**, such as the wheel of bliss shown here.

Aspects for Contemplation

The mandala (**large circle**) represents both chaon and universe (spacetime). The divine energy is divided into continua of the four dualities of force manifestation: attraction (light areas)-repulsion (dark), space-time, strong-weak, and electric-magnetic.

The **central dot** represents the smallest a chaon can be — the lowest energy at the coldest temperature , just above absolute zero in deep, intergalactic space free of bound chaons. The **innermost circle** around the dot represents the biggest a chaon can be and still project repulsive force, which continuously diminishes with distance.

The **middle circle** represents the farthest reach of pure repulsive force (**black spikes**), just beyond which chaons can "flip" and bind for the briefest periods to propel TTT (the electromagnetic "photon"), and sustain the most fleeting of compound particle, nuclear, and atomic attachments. It is also the point where time begins and can be stolen.

The **outer circle** (achievement of full light) represents perfect order (complete attraction, no repulsion; anticipated in concepts of eidos, heaven, nirvana, dharma, etc.), never achieved in the temporal plane.

The **outer circle** also represents the physical limits of chaonic (and universal) expansion, chaons being the only particles that can directly absorb, carry, and discharge the divine energy, expanding and contracting in size as they do, thus giving spacetime its peculiar elastic , fluid, and gelatinous qualities through chaonic power-sharing, or equitability.

There are *two kinds of chaons*: *Free Chaons* and *Bound Chaons*. *Free chaons are the ones that make up the 3-D fabric of spacetime*. They have mass and a tiny, omnidirectional, repulsive force, tempered by increasing affinities at a distance, and closer ones if coerced. Imagine them as ornery hillbillies of old, each just beyond shooting range of the next.

Bound chaons are what happens when extra energy hits them. They swell up, flip, and eventually become attractive, and begin to stick together in various numbers and configurations, and for varying amounts of time, forming compound particles. These particles, in turn, form bigger compound particles if hit with more energy, and so on up the line, until the *particle aggregations become the biggest and longest-lived ones, protons and neutrons*, the constituents of the atomic nucleus.

Many tiny configurations of bound chaons carry various forms of radiant and other motile (or kinetic) energy from one point to another. In the case of electromagnetism, they operate through a process called *Temporarily Transformative Transport*, or *TTT*. What happens is that each particular configuration carries its energy a specific short distance before simultaneously breaking down and transferring it to the next batch of free chaons, creating a new particle with slightly altered configuration that carries the energy forward, etc. This is why there is a *speed of light*. It *is actually the speed of TTT racing through the spacetime matrix*, temporarily collapsing time while stealing space as it goes. It is also the reason why *scientists are confused as to whether light and other radiant energy are a particle, or a wave, or maybe a vibrating string*. It is all of these, the *wavelength being the distance traveled by each TTT iteration*, and the *vibration coming from chaonic timekeeping and the transformation and disintegration cycles of the temporary (very short-lived), chaonic compound energy-carrying particles involved* — whose ever-changing configurations in flight encompass far more types than "photons," the "periodic chart" of energy particles, or even the whole known particle zoo.

That is because in the course of moving all those tiny bits of mass and energy around, there is a *chaonic toll*, a tiny mass and energy tax for making that energy move stuff this way in the first place. *A variable number of new free chaons are created by each TTT iteration*, depending on who the carrier is, because the carriers are being reduced by the energy expended to move them. These *newly liberated chaons* (bound chaons, after all, just wanna be free!) *are added to the population of*

free chaons in the spacetime matrix, expanding it. The more radiant and other kinetic energy being emitted and expended, the faster the expansion.

Protons and neutrons are the biggest, longest-lasting composited particles. They can last for billions of years. *Electrons* are big too, the biggest, longest-lived, most motile energy-carrying compound particles. Electrons are chaon trolleys — "cloud" aggregators and launch pads for all that radiant TTT activity (if you're a chaon, you can hop on an electron alone or in bunches, but you'll always hop off together in bigger, more organized and energized bunches called *photons*, at least for awhile!). These three big, Methuselah megaparticles form the *constituents of atoms. Atoms, of course, form molecules* and the world we know, or think we do.

"Black Hole" (p. 12) is a terrible misnomer, reflecting deep Human fears of the all-consuming monster, Chaos itself. These things should just be called *Dark Stars* (their original name), as they all started out as core cancer in the biggest regular lit stars, at first extending their lives with fresh hydrogen, but then abruptly replacing their hosts, often with a *supernova megafusion bang* that showers space with heavier elements (p. 14). *A black hole is the opposite of empty*. It is the heaviest and fastest rotating nit (twirling faster and faster in slower and slower motion, the farther in you go, as time deflates, or rather gets crumpled in by free chaonic pressure) in the plangent fabric of spacetime, stronger (and far denser) than the strongest star, and full of everything needed to build a galaxy, from densely packed free chaons to hydrogen nuclei. When mature black holes get too full, *they open vents and blow out jets of newly composited particles from their poles*, ranging from no-names to neutrinos, neutrons, protons, electrons, and even whole starter nuclei that emerge at the end, along with gamma, radio, and the other kinds of ray stream (pp. 12, 15). Spewing that created internal stuff all over space, around the black hole and beyond, seeds the next generations of big and small stars, including ones that will end as new black holes, and get eaten by each other and ultimately by the biggest black hole.

That's because *every galaxy* — except the most early-forming, rudimentary, disorganized ones (star clusters) — *is a gigantic, revolving free chaon storm*. An antihurricane, but on a huge scale. A *push* storm instead of a *pull* storm, a chaonic supergrye, built of the layered merger, accretion, and organization of smaller such storms, all the way down to our puny solar system. And the biggest of them, ellipticals (p. 12) or majestic galaxies like ours with spiraling arms, is a long-lasting gravity super-

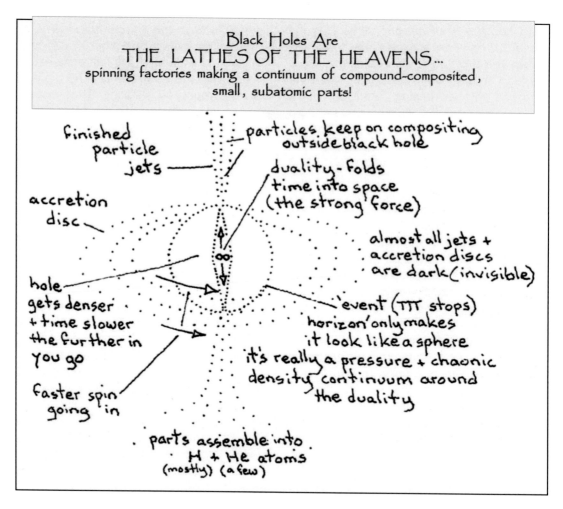

Black Holes Are
THE LATHES OF THE HEAVENS...
spinning factories making a continuum of compound-composited, small, subatomic parts!

finished particle jets —

particles keep on compositing outside black hole

duality-folds time into space (the strong force)

accretion disc —

almost all jets + accretion discs are dark (invisible)

hole gets denser + time slower the further in you go

'event (TTT stops) horizon' only makes it look like a sphere

faster spin going in

it's really a pressure + chaotic density continuum around the duality

parts assemble into H + He atoms (mostly) (a few)

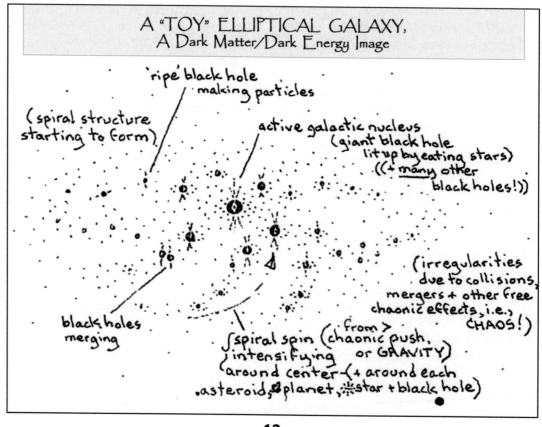

A "TOY" ELLIPTICAL GALAXY,
A Dark Matter/Dark Energy Image

'ripe' black hole making particles

(spiral structure starting to form)

active galactic nucleus (giant black hole lit up by eating stars) ((+ many other black holes!))

(irregularities due to collisions, mergers + other free chaotic effects, i.e., CHAOS!)

black holes merging

spiral spin (chaotic push, from > intensifying or GRAVITY) around center (+ around each asteroid, planet, star + black hole) storm with a supermassive black hole at its center, instead of an empty eye. A whirling, high-pressure, composite particle factory instead of a low pressure void. The biggest, most dynamically active thing in the Universe. All of this is powered by ornery free chaons pushed too close together, but not close enough to become friends for life, except deep in a big black hole. All this spinning, warping, and compressing of the fabric or matrix of spacetime and its imprisoned, bound-chaon bodies like stars, planets, and asteroids, is called — that's right — *gravity*. Gravity is pissed-off free chaons. And we all know what happens when you rile up the free (at least initially): Chaos!

Every single piece of composited matter, from the biggest heavenly objects down to individual atoms and subatomic particles, displaces free chaons, and takes their place in the fabric of spacetime. Warps their nicely-spaced layout, and pushes them closer together (*below*). So they do what's natural — push back. They try to take back what was theirs. Their space. That's what gravity does, especially on the practically invisible scale of subatomic particles, atoms, and molecules. And at that scale, they ultimately win, because many of these little composite assemblages, including complex organic molecules, are held together in places by relatively weak energy forces and bonds. By attacking, altering, and breaking these bonds over time, and interfering with subatomic, atomic, and molecular reactions, interactions, and other processes as well, *free chaons cause entropy, aging, decline, disease, dysfunction, despair, deterioration, disintegration, and death, not just of us, or all life, but of the inanimate things all around us, as well*. What wild creatures large and small, accidents, bad chemistry, fire, radiation of all kinds, water, and weather can't do, displaced angry mobs of free chaons accomplish in the end. Free chaons put the temporary into the temporal plane. And they put the free will (and free whim) in too, for good and for ill. That includes all forms of violent action, which by its very nature is chaotic, even when it comes from seeming order. And short of violence, the continuous pressure of free chaon attack on organized existence renders all of it increasingly structurally and functionally suboptimal for life. This is why everything in life is imperfect (including us), and never achieves its full potential. It's also why complexity has its limits.

So *there is an ecology of outer as well as inner space, and continua of cosmic manu-facture. Black holes are the hot presses* (a lot hotter than stars in their middles — pressure generates heat!), *and stars the finishing forges.*

Black holes (pp. 11-12) squeeze out the small parts needed to make long-lasting atoms and the gasses that become stars and planets; *stars* (p. 14) build up the starter atoms into all the planetary elements we know, and generate radiant energy. Inside each, along energy gradients mediated by pressure, temperature, and process (nuclear fusion in stars, the strong force in black holes) there are interwoven, flowing, cycling concentric layers or economic zones devoted to making specific atomic parts and elements. At the black hole's center there is not a singularity, but instead a *duality* that separates time from space, and folds pure time into what becomes time- and age-resistant ("timeless") electrons, protons, and neutrons. And the pressure created by free chaons displaced by bound ones (*a.k.a. gravity*) *organizes everything*: the star systems into massive composited dark star-centered galaxies; the conversion of big stars into starter black holes as second lives (starting as metamorphic celestial parasitoidism, and often ending in the spectacular stellar explosions call *supernovae*, which make the heavier elements); the compositing of black holes into bigger and bigger ones, with greater particle-creating capacities after each merger; culminating with the biggest things in space, the supermassive, hyperfast-rotating, particle jet-spewing dark stars (brilliantly illuminated when big black holes "pig out" on lit stars and gas clouds, and end up choking on too much matter) that anchor the largest galactic centers. *Gravity, the master force,* also breaks down everything made of bound chaons and recycles it in spacetime. *All is chaonic*.

৪৩೧೮

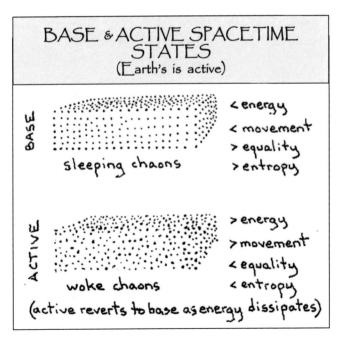

BASE & ACTIVE SPACETIME STATES
(Earth's is active)

sleeping chaons
< energy
< movement
> equality
> entropy

woke chaons
> energy
> movement
< equality
< entropy

(active reverts to base as energy dissipates)

13

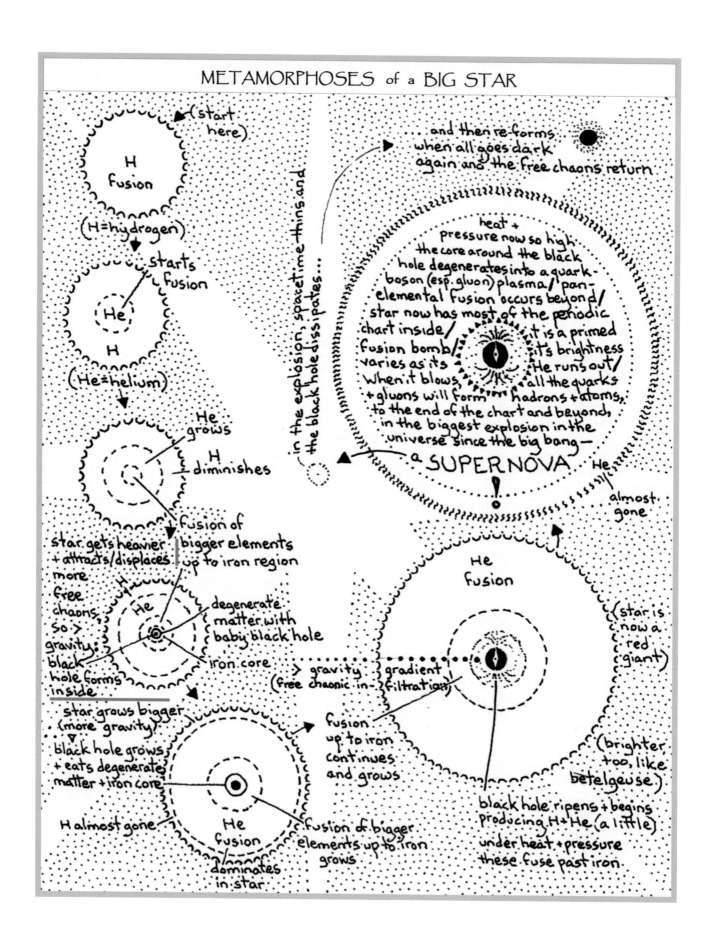

(start here)

H fusion

(H = hydrogen)

starts fusion

He

H

(He = helium)

He grows

H diminishes

fusion of bigger elements up to iron region

star gets heavier + attracts/displaces more free chaons, so >

gravity: black hole forms inside

He

degenerate matter with baby black hole

iron core

star grows bigger (more gravity)

black hole grows + eats degenerate matter + iron core

H almost gone

He fusion dominates in star

fusion of bigger elements up to iron grows

gravity gradient (free chaonic in-filtration)

fusion up to iron continues and grows

He fusion

black hole ripens + begins producing H + He (a little) under heat + pressure these fuse past iron.

(brighter too, like betelgeuse)

(star is now a red giant)

He almost gone

...in the explosion, spacetime thins and the black hole dissipates...

and then re-forms when all goes dark again and the free chaons return

heat + pressure now so high the core around the black hole degenerates into a quark-boson (esp. gluon) plasma / pan-elemental fusion occurs beyond / star now has most of the periodic chart inside / it is a primed fusion bomb / it's brightness varies as its He runs out / When it blows, all the quarks + gluons will form hadrons + atoms, to the end of the chart and beyond, in the biggest explosion in the universe since the big bang —

a SUPERNOVA !

He

He almost gone

PRELUDE TO CONVOLUTION

BUT IT DOESN'T END THERE. If we can tear you away from the existential horrors of contemplating what all those free chaons are doing to our insides (yes, gravity is also *inside* us!) and outsides right now, let's go back to the big picture one more time. There is a second reason that life is short. And there is surely a God, though not exactly the nodal one(s) Humanity has envisioned throughout history. A better one, who created the divine energy that is the Universe, for a purpose. And getting better all the time, as a result.

Using mock-continuum thinking and chaon theory, we can go back and re-imagine the history and evolution of the Universe, and go forward to its end (p. 17).

Energy-wise, the **Universe is like a big battery**, but it started materially as a-meadow-in-a-can. Inside that singularity (and there was only ever one) were all the makings necessary for what was to come. Just as seeds and nuts are dense, power-packed little things, so too all the chaons were bound at first by all that potential and released kinetic energy, in forms we've come to know and some we'll never know. Once the initial, glowy inflation slowed, and went dark, and star formation began in earnest, an inner void opened from **Big Bang** momentum (p. 17), and it became clear what the mature Universe would look like — a hollow ball. Actually, more like **a growing soap bubble membrane**, stretched thinner and thinner by all the free chaons released by frenzied kinetic energy and rapid cooling that went along with that first big expansion. And what was inside that expanding bubble? Nothing. Not even a chaon. Outside, too. No, wait. *Whoever* started all this was still on the inside, surrounded by **Hris** (that's a gender-inclusive, possessive, pronominal adjective) unfolding Creation.

As the first stars made by those freed chaons winked on and produced lots more free chaons, they were herded into galaxies through the magic of **relativistic gravity**, and **black holes** were born.

MAIA STAR-THROWER, MIDWIFE OF THE UNIVERSE
[Also see frontispiece & p. 39.]

The holes (*a.k.a.* dark stars) could, by merging and organizing and getting bigger and bigger as their surrounding galaxies grew, semipermanently bind chaons, producing the rest of what was released out of the original singularity; so the Universe was now made whole, and entered its mature, or **Maia**, state (*above*). It could fully reproduce. And through the magic of free and bound chaonic interaction, it could recycle. The meadow's ecology was now in summer.

The membrane of spacetime began to expand again, faster and faster. Stars and black holes/dark stars went through their own symcosmotic evolutionary cycles of birth, growth, maturity, decay, transformation, and merger as their increased numbers produced ever more and diverse matter. New galaxies formed and grew by free chaon-driven accretion around dark star solar systems and the merger of such systems. Radiant energy could transmit by TTT within the membrane (until it hit the curved edges, giving the impression of a disc-shaped Universe), but not across the inner or outer voids. When it got to the borders of spacetime, it died for lack of free chaons, a form of energy loss. We (our

15

star, our planet, life, and us) came about in this *Maia*, or mother, state of the Universe. We're still in it — for now.

If one looks at Einstein's most simple and famous equation as the

$$E = mc^2$$

summary description of the relationship between free and bound chaons, several things become clear. One is that he did not carry relativity far enough, or rather the implications inherent in those far more complicated equations. This was because he did not believe in an expanding Universe (the root, by his own admission, of his "biggest blunder"). Another is that as the spacetime membrane expands and thins, the total amount of energy is dispersed over a larger and larger Universe, a second form of energy loss. And that amount is diminishing by the unseen spiritual work that the kinetic part is doing. Nothing is recycled or transferred without these three forms of energy loss happening. Losses so tiny, close up, that they can't be measured. But in universal aggregate, losses which will show up more and more, as *Maia* approaches menopause (p. 39).

So *everything is relative*, even the speed, intensity, and direction of light, which varies, based on the varying passage of time, the density of the free chaonic matrix through which it passes, and its shape (curvature) and ceaseless motion. This means the immutable cosmos we think we see is a shimmering mirage of fossil light, whose sources are long and longer relocated or vanished altogether. That part is accounted for in Einstein's equations, because a warping, thickening, and moving spacetime matrix produces observable phenomena like gravitational lensing and the "event horizon" (actually, just a layer where free chaons get dense enough to block long light waves first, then shorter ones, deeper in) of black holes, where TTT stops working. And *if the speed of light is relative, so is time*, obviously, as part of the accelerating expansion of spacetime, and its local opposite, especially in and around massive objects like galaxies, black holes, stars, and even little planets like ours.

The bottom line is that there is *universal time inflation*. There's also *local time deflation*, near and especially in big stars and black holes, but it never reaches *absolute time zero* (*i.e.*, eternity), just like deep space never reaches *absolute temperature zero* until you get to the voids, where there are no chaons at all. (In the inner void, though, there is the consolation prize of increasing spiritual affinities, as one journeys to the center of *absolute love*, which more than makes up for lack

of temporal heat.) Time gets closest to stopping in the center of the biggest black holes; but *absolute pressure* (the only thing other than absolute zero that could stop time, and its duality, space, in the temporal realm) is never reached (neither is *absolute spin* in dark stars), because there are safety valves at both poles to blow off excess energy (of course, banked energy, which is what *matter* really is, in the form of all those compound particles created within). The safety valves put a limit on the sizes of the biggest black holes, and reflect the free chaonic limits (remember, they're ornery!) placed not only on supermassive dark stars, but on the sizes and complexities of the galaxies around them, and the stars and planets composing those galaxies.

Time inflation is just like money inflation. That feeling us oldies get of time going faster and faster, and leaving us behind, is real! If space is inflating, so is time, as they're joined into one by the matrix of free chaons forming the actual fabric of expanding spacetime, not just the mathematical one. And this expansion and time inflation is accelerating faster and faster; *i.e.*, on an ever-shifting exponential or logarithmic scale. We are hurtling towards an end because this dissipation by spreading is not sustainable. At some point the soap bubble is going to pop. Or more likely, just fade away, as the Big Bang battery runs out of energy from doing all that work, leaving little galactic islands that will later break up and fade away themselves. A history and fate both reflected and prefigured in works like Arthur C. Clarke's *Childhood's End*. Or T. S. Eliot's *Waste Land*, Algernon Blackwood's *Willows*, George Lucas' *Force* stories, Jung's *Red Book*. And of course the beloved works of Tolkien and the ancient Hellenic poets, philosophers, and myth-makers. In fact, all storytelling contains traces of the larger narrative of the Universe and our little planet.

Soulprescience, or spiritual awareness and yearning, is *the sixth sense of Humanity*. It, often subconsciously or through dreams, propels all art, all fiction, all attempts at religion: our inborn, Nature-driven sense (once powered by the divine terror of awe at the sublime; now horribly dulled by the technology accompanying and enabling the false abandonment and destruction of our Earthly mother) of the "Other-world," the spirit-realm,-plane, and -sphere that surrounds and awaits us and all life. *Every living thing has some degree of sentience*, and the shades of all — even little weeds, rodents, germs, and mosquitoes — have a likely home in heaven, where they can ultimately

16

become reintegrated with God, or perhaps starting off in the outer void, or purgatory, the movable feast where spiritual reintegration starts, as the bubble of Creation expands into and past it with the passage of spacetime. But unfortunately, we Humans are misled on matters moral and spiritual. This is the root of the fear and outrage that fuels internecine conflict capable of rising to war, along with want.

The good news for now is, we're in a bit of our own spacetime buffer here, on an arm of a big galaxy. Because we're in an atmospheric Biosphere on a free chaon-powered, gravity-driven planet, in a gravity-driven solar system, in a massive gravity-driven chaon storm (the Milky Way Galaxy), the full effects of intergalactic spacetime expansion are muted by the denser free chaonic matrices around us. That includes time inflation, which really means, nonetheless, that there is less and less universal background time available to do stuff each second, minute, hour, day, and year. That's a *huge* selection pressure on all living things. It, and the paradox of growing numbers of galactic free chaons around and in us, guarantee that individual lives will be short, and get shorter. ℬℭ

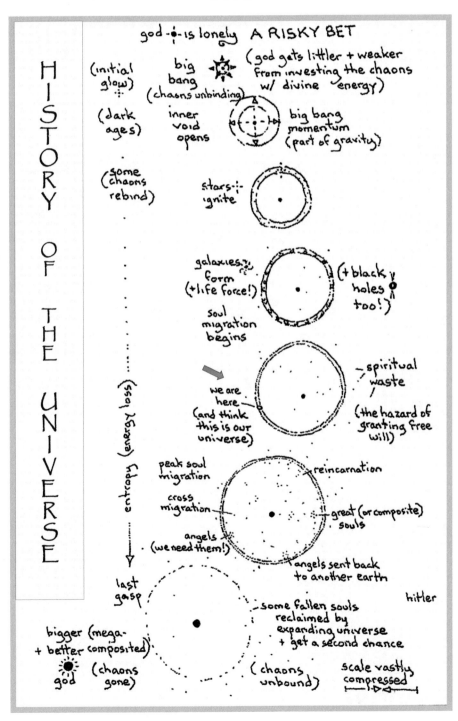

BUILDING BETTER BUCK MOTHS:
THE OTHER HALF OF EVOLUTION

THE PERIOD FROM **1903 TO 1918** saw the airplane go from the flimsy Wright Flyer of Kitty Hawk, all the way to the Fokker D-VII, a warplane so fearsome, in the right hands, that it became a mainstay of nascent air forces around the world, long after the Great War had finally been called off. This progress happened with increasing swiftness in under fifteen years, the last four under the intense selection pressure of aerial jousting among the famous "Knights of the Air" (p. 19).

What drove this process? Certainly inventiveness on the part of early aircraft makers like Fokker, Sopwith, Nieuport, Avro, S.P.A.D., and the rest. All had lineages of models, showing steady improvements layered and integrated one over the other, increasing in tempo until Armistice Day. And the aviators got better, along with their steeds.

But there was also ... *stealing!* Or, more euphemistically, *requisition.* Blander yet: *transfer.* Planes got shot down intact behind enemy lines, but their pilots couldn't light them on fire (they were all required to). Or Mata Hari came home with plans. Or someone with big binoculars actually spied something about a plane in the air. Sometimes by the next day, opposing planes had been fitted with the new development(s). Many turned out to provide only small benefits, some none at all. And a few actually caused harm. But some were big.

There is an apocryphal story of EDDIE RICKEN-BACKER, the greatest American ace of the war, one day pinning what was then a major development in the also young sport of auto racing (his former profession) to a strut next to his cockpit, then proceeding to shoot down five enemy planes. The invention that gave him a decisive advantage? The *rear-view mirror.* So the improvements sometimes came from sources far removed (but not too far) from aircraft. There is truth behind this tale, as much of the rapidly increasing horsepower in planes back then came from modified car engines, which had a head start on development.

As there is nothing new under the sun, Nature long preceded Humans in this practice.

A lifetime of study by any naturalist will confirm that ***everything in nature is marvelously adapted.*** Certainly this is true of the Buck Moths, in ways more spectacular and visible than in most forms of littler life. For example, what other moth rocks itself to sleep, then does it again if disturbed? And if messed with further, flips down into the duff, like a curly reversed umbrella, in full heraldic, aposematic display (p. 19)? Flies diversionary cover over the rest of the flight, when out of eggs or spermathecae, to take attacks by birds and teach them that Buck Moths taste bad? Or has dazzle-patterned caterpillars, exactly matching dappled sunlit habitats, which can vanish like tiny battleships in plain sight of tinier enemies, when not reminding the bigger ones that they are poisonous? And these are only a few; we can easily reel off ten times more.

Well, might come the answer, that's all due to

[Also see p. 40.]

The Last Dogfight

natural selection. But with all due respect, this is *the negative half of evolution* — the destructive half. Survival of the fittest. Darwin only had half a theory. *What causes the creative half that produces the fittest?* And what has consistently driven evolutionary progress in the capacity and complexity of organisms at an accelerating pace over time, despite the severe disruptive setbacks of mass extinctions?

The answer for a century has been variation in genes caused by "random" mutations. But this can provide only small, incremental benefits at best, and at a slow pace. Something else is at work. What is it?

A male Buck Moth's aposematic display
[Also see p. 42.]

There are clues. Hints, even. Some of them big. One of these is largely invisible to us: the **Cellosphere**, that mighty series of concentric, nested orbs filling the skies, seas, and Earth with what we dismiss as the most primitive organisms, *the ones with only one cell*. We even call the most primal of all **Archaea**. Yet unicellular beings account for the majority of life in both space and time. Always have. They were the only life for well over half of Earth's 4-billion-year history.

These tiny things aren't primitive at all. There's more subtlety, complexity, and sophistication in any one living cell than in a Boeing 787, or the fastest supercomputer, or even the entire Internet. The biggest reason: cells have far more continuum processes (cells *inside* cells!) going on than those nodal machines — exponentially more. We've been studying single cells for a long time, if for no other reason than that many of them kill us before the chaons can — and at the same time, we *multicells* are utterly *dependent* on them for our lives — but we still don't understand more than a fraction of what they do or how they do it. And we're *made* of them.

Flight as a metaphor for life turns out to have a lot of utility. Both are precarious, and governed by that megadimensional matrix of

19

interacting continua which represents reality. Both involve sailing vessels through an unforgiving environment. But **to master flight, Humans only had to understand six sets of interacting continua: lift, weight, thrust, drag, balance, and control**. The WRIGHTS were the first to do this, through study (of birds!), gap analysis (of poor LILIENTHAL and other doomed pioneers), modeling (one of the first wind tunnels), and careful and persistent trial and error. Everyone else who followed built on that (by *stealing*, according to the Wrights).

In contrast, life has to master pretty much all of the interacting continua there are on this planet, and then a whole lot more. That's because in addition to getting the sweet spots in the physics and chemistry departments right, life invented a universe of new ones by creating first, self-replication; then diversity, ecology, and a Biosphere that dwells within, and alters, the inanimate continua. It also alters itself. It evolves, just as the Universe does, in sync with the ecology and evolution of spacetime. And it creates its own stable nest: **Gaia**.

If one lines up the history of flight with the history of life, there are a lot of parallels that point to common process. The most important has precedent in the unfolding and self-construction of the **Maia**, or mature, phase of the Universe (p. 15). And it has a name, not much used in science (MALTHUS and his followers being the outstanding exception), that was coined long ago by *bankers*, who stole this process for their own purposes: **compound interest**. Compound interest is a specific form of **Compound Compositing**. Compound Compositing is Like Money in the Bank! It grows — faster and faster. What EDDIE RICKENBACKER was doing was building a better S.P.A.D. Through *stealing* — oops! — um, *transfer*, everyone who wanted one soon had a rear-view mirror. That went for cars, too, ultimately by diktat, like seat belts later.

Returning to planes, there first was a **fluxus** (p. 22), a period when some aircraft had it, some didn't. Meanwhile, other innovations were being tried, willy nilly. The **Aeroplane fluxus** grew and grew from the Wrights into the years of the Great War. The winnowing by the intense selection of aerial combat, and the intrinsic dangers of flawed early airplane designs, went on throughout. By war's end, the SE-5, S.P.A.D., and Sopwith Camel were flying in great numbers for the Allies; but the Germans, out of money, couldn't build enough Fokker D-VIIs to keep up, and had to rely on old Albatros (the famous Triplane had washed out with the Red Baron). And then the *biplane era* faded into a new lineage of *metal monoplanes*; and the race began anew, pro-

ducing the even bigger fluxus of *WW-II aircraft*. And that one again birthed a new lineage, *jets*, and most propeller planes in the years after became *ghosts*. The beat goes on today, with *helicopters, drones, metamorphic Top Gun jets, VTOL, stealth*, etc. Facilitated, of course, by plenty of *stealing*, um, *transfer*, back and forth. The annals of history are replete with such stories, starting with the acquisition of fire and invention of the wheel. Their collective moral is that state secrets, trade secrets, personal secrets — really, secrets of any sort — don't last long. (Plagiarism rules genetics in a transfer paradox: **the mirror of stealing is giving**.) Along the way, rear-view mirrors in planes turned into *radar*, and then *GPS*. Promising new lineages in aircraft, like the SST, have also come and gone, due to familiar extinction forces (being too energy-expensive!).

So what we saw in Buck Moths was basically the same thing: a **fluxus of varyingly fertile moths with different degrees of hybrid ancestry that started with re-contact by two supposed species (close relatives, obviously, but different in important ways) that produced a new lineage — our new moth**. Something related, but different. Something that was better adapted for the immediate postglacial environment at hand. Better able to exploit it. And the process that produced this new and (formerly) more fit thing was — **Convolution!**

Convolution is the Solution to the Problems with Evolution. It is the **key creative force of evolution,** and **the cause of evolutionary progress in complexity, diversity, and overall (general) fitness of organisms. It solves the Species Problem**: At all times, lineages are being improved through the addition of new **layered adaptations**, which are the products of **compound compositing**, driven by **Convolution Events,** big and small. **There are NO SPECIES** (the ever-widening **Extinction Gaps** between many surviving lines make them *seem* like species), just lineages of related and less related **Semispecies**, crossing and re-crossing at varying intervals, like braided strands of rope (p. 35). Sometimes they produce new lineages, sometimes they die, sometimes they fade into something else, and sometimes they bend to meet other, more distant lines. **Genes are the money of compound compositing**: fungible, like Lego blocks. They can and do flow between and among lineages, including distant ones, using different mechanisms. The Cellosphere constantly delivers genes to multicellular organisms in a set of processes collectively called **Genetic Bombardment**. But genes also fit and work together in certain and varying ways, among themselves, their helper molecules, the bodies they are in, and the physical, chemical, and biological environment, including all other organisms, outside and inside the soma, the vessel. They do it in a manner best pictured in reduction as the **Epigenetic Triangle** (p. 9), which varies for each lineage, and in the **Metamorphic Varieties** produced by interactions among genome, soma, and environment. **Compound compositing** produces **Layered Adaptations** that build up the **General Fitness** of all organisms to weather shifts in the megadimensional matrix of interactive continua, which control the physical, chemical, and biological aspects of inanimate and animate existence. And through layered levels of feedback and control loops involving all the elements of the **Epigenetic Trinity** (genome, soma, and environment, p. 9), it widens **Metamorphic Amplitude** to respond to a wider and wider range of **Expression Triggers** (the "Chitty Chitty Bang Bang" effect).

Free Chaons intersect and interact with Convolution in several vitally important ways. First, they **alter and age all organisms, and limit lifespans** by pushing on subatomic, atomic, and molecular bonds and reactions. In so doing, they cause by far the greatest number of genetic mutations, and genomic shifts, at a rate far faster than all other causes combined. They are the major cause of so-called **genetic drift**, and **in-soma mosaicism**, among many other metamorphic phenomena. Second, **by forming the matrix of spacetime, and being the agents of its accelerating expansion, they cause enormous, varying, and increasing selection pressure on all of life through** Time Inflation (all organisms are products of their own time). This forms a second, absolute limit on the lifespans of individuals as well as lineages.

For example, **we'll never get the dinosaurs back;** doubly so because our gravity has become stronger since then, as our galaxy has grown in free chaonic density. You couldn't loft a pterosaur the size of Pegasus on Earth today, nor could dinosaurs as big as whales live on land now. Triply so, as there also was more in-air oxygen back then. And finally, there's the issue of orbital slip-back — Earth was closer to the sun in the Age of Dinosaurs. That's why the Ice Ages only happened later, when we came along. And don't forget so many other shifting variables sliding along interacting continua over long time periods, like sun intensity, Earth tilt, orbital shape, day length, wobble, moon distance, etc. So sorry, no more giant horsetails, clubmosses, or dragonflies ever again, either, just their mini-descendants.

But there's a good side to time inflation, as well. It makes succeeding generations shorter and short-er, in each defined time interval, say a year, or one rotation of the Earth around the sun, because the time intervals in absolute terms are shrinking with the passage of accelerating time. At the same time, the free chaonic spacetime matrix is densifying locally, as our galaxy grows and spews out more and more chaons. So more mutations, more Convolution, and more overall evolution happens as the intervals and iterations pass by faster and faster. Speeded-up evolution! On Earth and in space. That explains why the history of life is back-end loaded, and so therefore must be the history of the entire Universe.

Not only that, but some groups, like insects, have not just kept up, but have overcompensated with the faster running of time and strengthening of local gravity. They've learned to do more with less. Their amazing fecundity and metamorphic amplitude is what put them ahead of the curve. And it's why they came to dominate the terrestrial Earth, with their diverse, semi-miniature hordes (and getting smaller than they used to be, like everything else), until now.

Contrast the insect victors of the present with the slowly reproducing dinosaurs of yore, who "thought" they had all the time in the world because they had a lot more than we do (and fewer killer chaons in and around them). Another reason they're gone, never to return.

Insects embody the magic of metamorphosis, but that's just the beginning. **Metamorphics** (p. 9) encompasses not just form, but function and behavior, i.e., all life processes in individual organisms, including, especially, **ontology**. And **all organisms behave**, some more slowly than others. This includes plants, fungi, and cells. That's because life molecules behave, through their symbiotic hooks. **Chemistry is at the core of symbiosis**; that's why Organic Chemistry was so hard!

The fact that the ancestral Cellosphere has been able to persist and adapt to the time inflation phenomenon (however buffered it may be by galactic gravity) by changing many of the dynamics of life itself, over time — never mind producing and sustaining an explosion of accelerating multicellular diversity now being brought low by the Sixth Mass Extinction caused by Humans — is truly miraculous, and stands as testimony to life's stubborn persistence in hazarding the perils of **the only seven truly universal laws**:

I. Time never runs backwards.
II. There is no Locality: All is in constant outward motion.
III. Everything is Relative: There are no additional fixed laws operating across all time, space, and

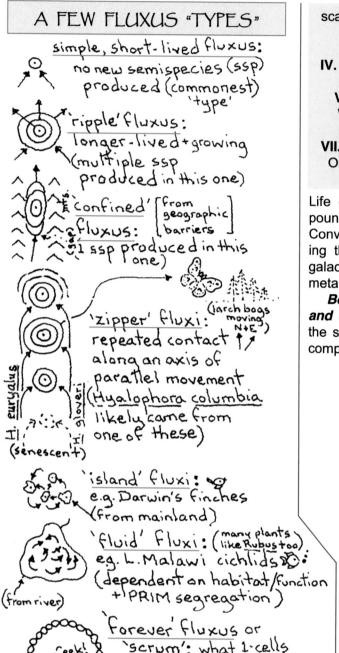

A FEW FLUXUS "TYPES"

simple, short-lived fluxus:
no new semispecies (ssp) produced (commonest) 'type'

'ripple' fluxus:
longer-lived + growing (multiple ssp produced in this one)

'confined' [from geographic barriers] fluxus:
(1 ssp produced in this one)

'zipper' fluxi:
repeated contact along an axis of parallel movement (*Hyalophora columbia* likely came from one of these)

(larch bogs moving N+E)

H. euryalus

H. slover

H. (senescent)

'island' fluxi:
e.g. Darwin's finches (from mainland)

'fluid' fluxi: (many plants like *Rubus* too)
e.g. L. Malawi cichlids:
(dependent on habitat/function + PRIM segregation)

(from river)

'forever' fluxus or 'scrum': what 1-cells are constantly doing (source of constantly new genetic bombardment + infiltration/infection of multi-cells)

eek!

LAST ONE·BACKDOOR CONVOLUTION
THE REST·FRONTDOOR CONVOLUTION

---

scales, just ever-shifting processes of continua and temporary nodes, including the abrupt discontinuities of chaotic events.

IV. The destructive action of Free Chaons limits the development of order and complexity.

V. Temporal space has only three dimensions.

VI. Time and Space are Force Manifestations, not dimensions.

VII. Chaonic Gravity creates the Master Paradox of Order and Chaos, and all the other paradoxes of temporal Nature that flow from it.

Life on Earth, through its master process of Compound Compositing using Symbiotic Hooks and Convolution, has met all of these challenges, including those from time inflation, buffered by increased galactic free chaonic densification, through growing metamorphic amplitude, until we came along.

Bound chaons, of course, are the stuff of life, and everything else we know. And the Universe is the soul farm of a constantly growing and improving, compound composite God.

ଛୠୠ

> *Hemileuca iroquois* females will lay eggs on any fibrous or woody stem near *Menyanthes*. Here one is laying on Three-way Sedge (*Dulichium arundinaceum*). Some females lay half of their eggs in their home fen, then disperse to find a new one (p. 41). They can fly at least ten miles with half a load.

RECONSTRUCTING LINNAEUS

SEXUAL REPRODUCTION EVOLVED to maximize generational genetic (and resulting organismic) variety upon which natural selection acts; and to produce and protect temporary lineages of multicellular life from the rapid genetic transfer scrum (p. 22; *Backdoor* or *Non-sexual Convolution*) — which is the lot of the vessel-limited (one-celled!) but necessary Cellosphere that birthed and still swaddles and nurses the intercellular symbioses which ultimately became such complex organisms. *Frontdoor* or *Sexual Convolution* exists to build up these multicellular lineages in complexity and resilience through the acquisition and integration of *layered adaptations* created (sometimes by backdoor Convolution) over long periods of lineage separation. *Both forms of Convolution result in Genetic Aggregation*, the concentration of new and varied template material over time for selection to hone into better-adapted organism lineages. *Reproductive isolating mechanisms* (RIMs) and cellular differentiation and specialization, built up through frontdoor (hybridizing Convolution events) and backdoor Convolution (genetic bombardment from the enveloping Cellosphere, which includes the gene-collecting, -editing, and -injecting Virosphere), are the *mechanisms that produce and protect lineages*. But all RIMs are actually layered, partial reproductive isolating mechanisms, or PRIMs. They can occasionally be defeated sequentially in time and circumstance, especially by the additional variation-creating, mutating action of free chaons.

All organized existence, especially life, is at constant war with the two-headed monster of free chaons and time inflation. This is the overarching, literally physical *Struggle for Existence*, not the relatively puny biological internecine competitions, depredations, and seeming conflicts among lineages, which looked at continuously, are *symbioses* (p. 9). The actions of free chaons and time inflation on life mean short individual lifespans, and less and less time for life to fulfill its essential processes and activities — thus requiring those processes and activities, and the lifeforms that can accomplish them, constantly to change, improve, and adapt from one generation to the next.

The production and maintenance of biological diversity relies on a dynamic tension between stability and change. Symbiotic hooks, and the lineages that carry them, *multiply and layer* in times and places of relative stability, to the point of exuberant, sometimes excessive and bizarre *niche luxuriance*, especially in the tropics and in relatively isolated spots like Hawaii or the Galapagos; but are *winnowed* in times and places of rapid change. Such winnowing is accelerating now, because of the rise and ever more disruptive activities of Humans.

THE RECORD OF BIOLOGICAL DIVERSITY *is encased in a nodal, hierarchical system* that was invented by LINNAEUS over 250 years ago. *The fundamental category in that system, the species, does not exist as a nodal entity because of Convolution*, and the megadimensional, interactive, multicontinuum nature of reality in general. This is the cause of the *Species Problem*. It threatens to take down the entire hierarchy, which is the organizing principle of biological knowledge.

We can preserve the Linnaean framework, which is an ongoing, elaborate attempt at an orderly historical projection into the past of life lineages and their divergences, extinctions, and extinction spans, or gaps, frozen in present time, while allowing for the growth of biological and other sciences from purely nodal into mock-continual concepts like Convolution. The solution for sexually reproducing organisms is simple: *replace the Species category with another, the Semispecies*, for identifiably separate lineages open to future frontdoor Convolution events. Semispecies are genetically Leaky Species, the lineage-based players of Convolution whose somas, or vessels, leak genes in and out across the entire span of higher taxa, like ships do water from any ocean.

The mechanism is simple, too: *downgrade all such lineages with species names to Semispecies*, and make the first- or only-named (monotypic) species in each group of closely related lineages the *Nominate Semispecies* of the group, instead (in Buck Moths, that's *Hemileuca maia*). This actually upgrades them in the eyes of evolution, *recognizing them as the active agents* of a dynamic process of variable forms of progressive change, including potential future backdoor and frontdoor Convolution events.

At the same time, *upgrade all named Subspecies that deserve it to Semispecies* (many do). The rest are *metamorphic varieties* and *simple polymorphisms within semispecies lineages*. Metamorphic varieties (and even important polymorphic ones) also should be recognized as such, cataloged, and, where needed, given names more memorable and meaningful than anonymous numbered or lettered placeholders. Latin is a beautiful language; one of its joys is that almost anything can be Latinized. In that way it's similar to English.

Conversion of the Linnaean taxonomic system at the species level will need to **vary among life forms**. The Platonic species concept, based on *Eidos*, never worked well for vessel-limited cells and viruses; and neither has the medieval *Great Chain of Being* for more complex life, which undergirded Linnaean hierarchy. They have been problematic in many and varied ways for plants, fungi, lichens, and numerous other higher, multicellular taxonomic groups in Linnaean classification. This is because varying Convolution processes exist among higher taxa, due to their divergent symbioses, life strategies, and methods of reproduction, and the intra- and extra-lineage variations produced, including many metamorphic varieties arising from single genomes that are triggered by epigenetic triangle effects, which run on multiple continua themselves. The taxonomic solution is to create categories of **metamorphic varieties**, rather than Semispecies, to contain this form of diversity.

An example in Lepidoptera: The small, pale spring form *canadensis* of the Eastern Tiger Swallowtail (of the *Papilio glaucus* group, Papilionidae, p. 25) may be a winter stressor-induced (by length, severity, onset, coldness, and drought accompaniment of a particular year), **seasonal metamorphic variety**, produced by the same genome that codes for the summer forms. Different layers of genes are activated, modulated, or suppressed to produce it. Global warming has exposed this by scrambling the seasonal cues in the epigenetic triangle of the Tiger Swallowtail so that the spring and summer forms recently emerged together in the Catskill Mountains of N. Y.

Another example in plants: Dwarf Pine (p. 25) is a relict **postglacial adaptive metamorphic variety** of Pitch Pine (*Pinus rigida*, Pinaceae), found in a handful of extreme pine barrens environments in the northeastern U.S. that are characterized by very droughty, fire-prone, poor, and strongly acidic soils. The same genome that produces tree-sized Pitch Pines makes them wizened shrubs under those environmental conditions by suppressing apical dominance. That suppression has been lifted by recent heavy rains brought on by **rapidly accelerating global warming** (**RAGW**; hotter air holds and releases more water), so that the former "Dwarf Pine Plains" of the N. J., Long Island, and Shawangunk Mountains (Ulster Co., N.Y.) Pine Barrens are turning into scraggly, tree-sized pine stands.

We immediately need to launch a worldwide effort to find and name unnamed, active Semispecies, distinguishing them in part by, and from, the metamorphic varieties and simple (so-called "balanced," often single-locus) polymorphisms that

are found in each **semispecies lineage**. They are the **repositories and agents of ecology and evolution, the machinery of biodiversity.** This will require the efforts of thousands of amateurs, trained quickly and guided by professional taxonomists, whose numbers must also increase, as there are many millions of such Semispecies waiting to be found. Geneticists, too, could save much time and effort by developing a fuller understanding of whole genomic dynamics (including variations in genomic epigenetic triangle interplay with soma and environment) and its uses in distinguishing Semispecies, metamorphic varieties, and polymorphic forms. Individual genes change their behaviors, functions, and productive outcomes as they move from their genetic, somatic, and environmental surroundings in one lineage to another, or even within lineages (jumping genes), so the erection of a new nodal species definition, the "genospecies," based merely on comparing genetic lists of lineages, or genomes, is misguided. The dynamics of Convolution and the true workings of biology (it's *all* symbioses) ensure that Semispecies are far more than mere sums of their genetic parts. In the face of the Sixth Mass Extinction being turbocharged by RAGW, there's no time to lose, fooling around with failed and misleading nodal taxonomic concepts.

Discerning and describing most unnamed Semispecies will be facilitated by the existing Linnaean classification framework. (Genera will also need to be adjusted to reflect geologically recent frontdoor Convolution spans among related Semispecies lineages.) Undescribed Semispecies exist mostly in lineages already given Latin names; in other words, they have already been recognized as species or subspecies, "sibling" groups, or closely related species. **The description can therefore be a stripped-down, diagnostic one**, listing the distinguishing differences between the new Semispecies and the nominate Semispecies in a (formerly known as) *species* lineage, and any other named Semispecies in that lineage group. The recent history in geologic time of each Semispecies group, including movements, contacts, and evidence of past Convolution events (not all will have active fluxi or fluxus remnants to point to; deeper epigenetic triangle-based evidence will need to be gathered), will have to be sussed out accurately before new names are applied. For new lineages without known siblings, the description will resemble the old form done for new *species*, as a diagnostic description cannot be written.

Active **fluxi should also be cataloged by type** (p. 22) and named. The names can be geographical, with parentheticals containing known-source Semispecies lineages. Fluxus types range along continua of development, history, decay, and reformation with regard to environmental factors, especially those producing geographic isolation and connection for varying times and at varying scales. Individual populations in fluxi should be identified as such with the fluxus name and a geographic name for each population.

We offer this first description of a Semispecies as an example. In this case, the new Semispecies arose from continent-wide movements leading to the re-contact of two long-separated older Semispecies lineages in a **_classic example of front-door Convolution_**, and we know what the parental lineages were. They have names, and now a history. We also know one benefit this Convolution event conferred on the new entity: increased winter cold-hardiness; another linked one was moisture tolerance; and a third, adaptation to alkaline soils. Only one of its parents, the western one, had them in relation to the other parent. That was important in the eastern, relatively near-ice, waterlogged, post-glacial environment into which this entity expanded, feeding on a new, very cold-hardy, wetland-inhabiting, and fen-flourishing starter host. (The eastern Laurentide Ice Sheet of the last North American glacial epoch, the Wisconsinan, was much bigger, driving farther south, and wasting away more slowly than its western Cordilleran counterpart, due in part to the presence of the smooth, hard, ancient granitic Canadian Shield that dominates northeastern North America; p. 5.) Our little moth now exhibits an almost futile set of adaptations, in the face of rapidly accelerating global warming.

℘ℭ

❯ ❯ **Dwarf Pine** (_Pinus rigida_) at the **_Long Island Dwarf Pine Plains_**, New York, in November 1980, prior to visible impacts of global warming. Its _serotinous_ (closed) cones require fire to open.

❮ A small, pale spring form of the **Eastern Tiger Swallowtail** (_Papilio glaucus_) in the **_Catskill high peaks area_**, New York, in summer 2017. Fallen Hemlock needles in the photo average 1/3 in. long.

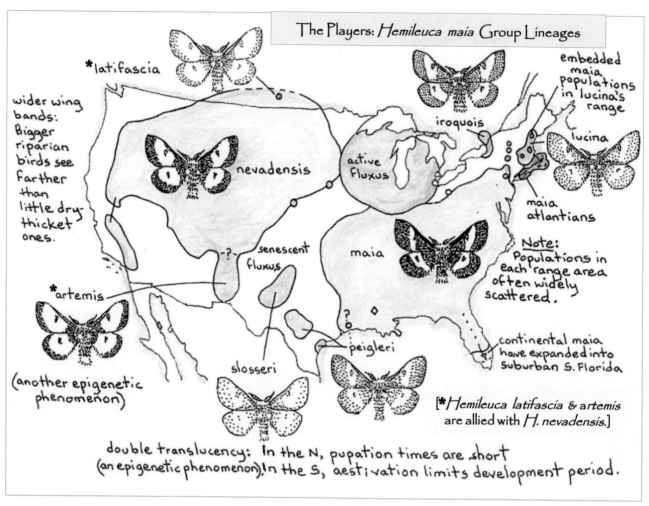

The Players: *Hemileuca maia* Group Lineages

*latifascia

wider wing bands: Bigger riparian birds see farther than little dry thicket ones.

nevadensis

active fluxus

iroquois

embedded maia populations in lucina's range

lucina

maia atlantians

Note: Populations in each range area often widely scattered.

?

senescent fluxus

maia

*artemis

(another epigenetic phenomenon)

slosseri

peigleri

?

continental maia have expanded into suburban S. Florida

[*Hemileuca latifascia & artemis are allied with *H. nevadensis*.]

double translucency: In the N, pupation times are short (an epigenetic phenomenon). In the S, aestivation limits development period.

Hemileuca iroquois (*H. maia* × *H. nevadensis*), S. n. p. [*Semispecies nova prima*]
[BOG BUCK MOTH]

Diagnostic Description: All life stages are similar to the eight named members of the *H. maia* semispecies group (*above*), with the following cumulative diagnostic differences:

Egg Ring: Tending to stay light greenish under heavy snow all winter; generally not bronzing from sun exposure (pp. 22 & 27, front & back covers).

Larva: Stays pure black to maturity; has a very limited, broken spiracular stripe in later instars, composed of scattered, small, creamy white dots (p. 27 & back cover).

Pupa: Morphologically similar to the others, but placed above the fall saturation line in floating fen mat matrices; sometimes in *Sphagnum* peat (p. 27

& back cover), more often in "grass" peat composed primarily of sedge and rush stems.

Adult: *White wing bands* (the best diagnostic character) are narrower than in *nevadensis*, *latifascia*, and *artemis*, but wider than in *maia*, *lucina*, *peigleri*, and *slosseri*. Medium **wing translucency** for the group — appearing less opaque than all but northern *lucina* and the northern part of the Great Lakes Buck Moth Fluxus in Michigan (*above*). Appearing equally as translucent or "glassy" as some *peigleri* and *slosseri*, but the translucence of *iroquois* is more even than in the Texas moths, which appear more peppered, especially as they wear. [S-to-N clines of increasing translucence occur in the ranges of *lucina*, *iroquois*, the Great Lakes *Hemileuca* Fluxus, and *nevadensis* (including *latifascia*).]

Overall Impression: One of the largest and most handsome of the group (pp. 3, 30), with a crisp, striking appearance, blending, in "missing link" style, the signature adult visual aspects (especially band width and opacity/translucence) of all three originally described Buck Moth lineages: *maia, lucina,* and *nevadensis* (pp. 3, 30).

Flight Pattern: Slower, lower, and weaker than the others, except when dispersing or alarmed, with less of the powerful milling, weaving, up-and-down motion typical of this group in level flight (pp. 40-42).

Habitat and Ecology: Found only in floating, alkaline fen matrices (pp. 27, 29, 41) of medium nutrients and pH (7.2-7.8), where it uses *Menyanthes trifoliata* exclusively as its **starter host** (p. 2 & back cover). Has the widest range of **finishing hosts** [over 100 species identified, mostly fen and mixed bog/fen inhabitants, but edge and upland plants as well, including willows (*Salix*), birches (*Betula*), and oaks (*Quercus*)]. Females lay egg rings on a wide variety of woody and winter-persistent herbaceous stems near *Menyanthes* (which dies down completely below saturation line for the winter).

Phenology: Eggs overwinter, hatching in May-June. Larvae feed into July, and pupate. Adults fly diurnally in late September and early October.

Range: Only three populations remain: two in southern Ontario Province, Canada, and one in New York State, U.S.A. All are located within the bed of former Glacial Lake Iroquois (pp. 5, 7, 26).

Status: All populations are currently listed and protected as endangered entities within their respective jurisdictions.

Conservation: A late postglacial, northern, relict semispecies lineage of the Buck Moths, reduced by habitat destruction following European and forced-African colonization to a handful of remnant fens. Now in dire danger of extinction by rapidly accelerating global warming (RAGW), especially from seasonal cue disruptions, combined with other

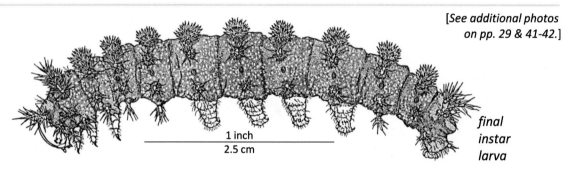

[See additional photos on pp. 29 & 41-42.]

1 inch
2.5 cm

final instar larva

Life History and Habitat of *Hemileuca iroquois*

empty egg ring

empty pupal shell & face cap

open fen habitat (type locality), 2017 [see p. 41]

Paratypes and Genitalia of *Hemileuca iroquois*

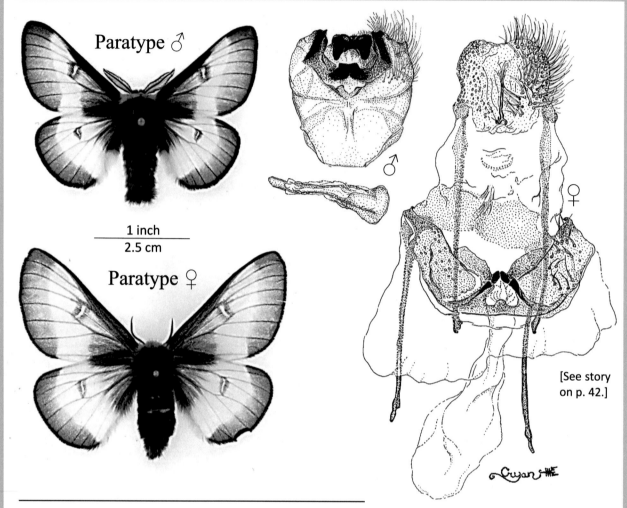

Paratype ♂

1 inch
2.5 cm

Paratype ♀

♂

♀

[See story on p. 42.]

Photos are at the same scale. Compare Elaine R. Hodges' drawings of ♂ genitalia of *H. maia, lucina,* and *nevadensis* on p. 118 of D.C. Ferguson's *Bombycoidea. Saturniidae, comprising subfamilies Citheroniinae, Hemileucinae (Part)* in THE MOTHS OF NORTH AMERICA, Fascicle 20.2A, 1971. The type series of *H. iroquois,* including the ♂ Holotype and ♀ Allotype, is at the Cornell University Insect Collection.

stressors, including rapid heating effects on fecundity and survivorship, and recent, record-high, persistent flooding of the fens in spring. Five of eight populations discovered in the 1970s-1980s are now extirpated, and no new ones have been found. This semispecies, as with the rest of the Buck Moths, is an **outcolonizing, semi-fugitive "base swarmer"** that depends on shifting mosaics of **seral conditions** (open fens) created by regular disturbance cycles (normal *flooding* in the case of *H. iroquois,* as opposed to *fire cycles* in the open-canopied or non-forested uplands used by *maia*).

Last-ditch efforts to save remaining lineage populations might focus on maximizing, restoring,

and maintaining open fen habitats that support abundant *Menyanthes*; also on **reintroductions**, including long-distance ones, into suitable, more northerly fens, if any can be found. Fen sites in Iceland, northern Europe, Siberia, western Canada, and Alaska could be considered for translocation as RAGW advances. [*H. lucina* (pp. 3, 30, 32) also is imperiled by RAGW, and might be introduced into Nova Scotia and other parts of the Canadian Maritimes, where Meadowsweet (*Spiraea alba,* var. *latifolia,* Rosaceae), its starter host, still thrives in bog edges, wet meadows, and old beaver pond sites.] Areas infested with Purple Loosestrife (which works as a starter host for reared *H. iroquois*) might be used as temporary stopgaps.

Finally, **long-term cryogenic preservation** of gametes and zygotes should be initiated, as part of a worldwide effort to save the millions of semispecies, which will otherwise succumb to RAGW (see p. 37). If Humanity comes to its senses and acts in time to stop and reverse RAGW, extirpated or extinct semispecies resuscitations and reintroductions into appropriate restored habitats can be done using this material.

Etymology: This semispecies is named for Glacial Lake Iroquois (p. 5), whose once-extensive wetlands secured its migration from the Great Lakes Buck Moth Fluxus (following re-contact between the older *maia* and *nevadensis* semispecies lineages in or near the Wisconsin Driftless Area), ca. 5,000-6,000 years ago. The name, by implication, also historically references the ancestors of the great HAUDENOSAUNEE People, who inhabited this region when the moths arrived (front & back covers).

Note on Saturniid Size: Distinguishing measurements of the types are not specified, but our photo scales show sizes of type specimens. Due to the effects of RAGW, many saturniid moths have responded metamorphically, for the time being, by becoming smaller. That includes this semispecies. Adult individuals at the type locality today average 5-15% smaller than those of the type series, which were collected approximately four decades ago.

Type Locality: Eastern shore of Lake Ontario, New York State, U.S.A. No further details are given to protect the type population, the only one known in this country.

Location of Types: The **type series** (HOLOTYPE ♂, ALLOTYPE ♀, and several PARATYPES) is at the Cornell University Insect Collection in Ithaca, N.Y. Additional PARATYPES are at the Canadian Museum of Nature in Ottawa, the New York State Museum at Albany, the American Museum of Natural History in New York City, the Smithsonian Institution in Washington, D.C., the McGuire Center for Lepidoptera and Biodiversity in Gainesville, Florida, and The Natural History Museum in London, England.

Botanical Vouchers: A few wild (hatched) egg rings of *iroquois*, collected on various pressed plant stems, as well as vouchers of the *Menyanthes* foodplant and general flora of its fen habitats, are at the Bailey Hortorium Herbarium at Cornell University in Ithaca, N.Y.

Suggestions on Semispecies Nomenclature: In naming new semispecies arising from **frontdoor** (hybridizing) Convolution, the originating semispecies (usually two, but possibly more) of the Convolution event which produced the new semispecies should be placed in parentheses after the new name (as above). If only one of the originating semispecies is known, that name should be placed in parentheses followed by "× second parental semispecies unknown" after the new name. If both parental semispecies are unknown, or if the new semispecies likely came from a **backdoor** (non-hybridizing) Convolution event, the name of the nominate semispecies of the group should be placed in parentheses following the new name, with the word "group" after it. If the new lineage is distinct from any other semispecies lineage group, the term "*Semispecies prosapia nova*" should follow the name.

John Cryan in a brushier habitat of *Hemileuca iroquois* on the east edge of Lake Ontario, on 21 September 1984. The trees are Tamarack and Red Maple. The white-capped stems are Tawny Cottongrass, emerging from the fen matrix of sedges. Buckbean, largely died down by this date, occurs in wetter areas. The moths fly throughout open expanses of the fen. [*See more habitat scenics on pp. 27 & 41.*]

The Four Northern North American *Hemileuca*

maia, Albany Pine Bush, N.Y.

"Atlantian" *maia,* Long Island, N.Y.

lucina, Belchertown, Massachusetts

lucina, Bowdoin, Maine

nevadensis, San Bernadino County,
California

iroquois, Lake Ontario
East Shore, N.Y.

1 inch
2.5 cm

Male specimens, all to the same scale. [Compare art on p. 3.]

Hemileuca Variations from the Great Lakes Fluxus

Webster Lake Fen, Indiana

1 inch
2.5 cm

Barry State Game Area, Michigan

Madison, Wisconsin
(abdomen removed for dissection)

Houghton Lake, Michigan

Wisconsin [*no details*]

Male specimens, all to the same scale

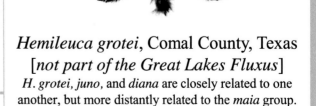

Hemileuca grotei, Comal County, Texas
[*not part of the Great Lakes Fluxus*]
H. grotei, juno, and *diana* are closely related to one
another, but more distantly related to the *maia* group.

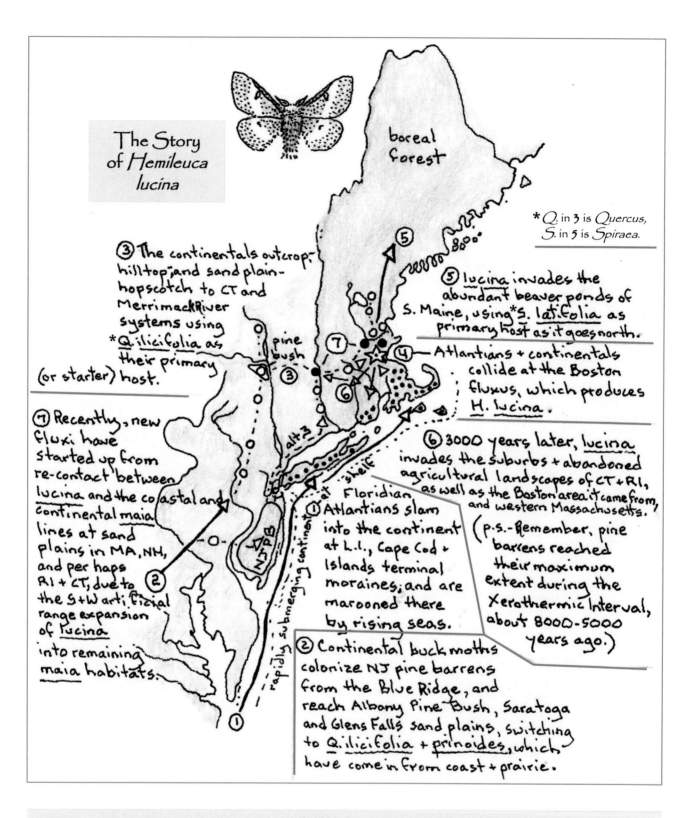

This complicated visual puzzle echoes efforts to understand *Hemileuca lucina*, underscoring the importance of integrating landforms, biogeography, evolutionary pathways, and deep knowledge of the moths. Please follow numbers 1-7 on paragraphs and the map to unravel the threads of the story.

CONCLUSION: A PLEA TO HUMANITY

THE AIM OF EVOLUTION IS MALLEABILITY: to stay at least one step ahead of the chaonic Reaper (free or bound), and by the accumulation of layered adaptations, drive forward more resilient lineages. Within the Epigenetic Triangle, free chaons simultaneously influence all parts of the matrix of layered feedback loop continua, comprising not only the genome, but all the other molecules, structures, and processes of the soma — as well as that of every organism inside (both microbiome and multicellular symbionts) and out with relations to it, and every aspect of the nonliving surrounds as well. Much of this interference causes winnowing of individuals; a small amount produces innovations by shifting the genome; and a tiny amount by shifting the entire epigenetic matrix itself, not just the genome. Whole matrix-shifting produces some of the most striking and discontinuous innovations, including crucial ones to keep up with time inflation. When combined with Convolution, this spread of chaonic effects can produce rapid leaps in far fewer generations than we thought. The Buck Moth creation story reveals how fast.

The genome is a nodal list of parts whose whole is far greater than its sum, because genes are malleable template parts that change functionally, as do their products, in relation to one another, to the parts and continuum processes of the soma, and to the continuous and shifting demands of the environment. In this regard, **Lamarckism** lives, primarily through **Metamorphics**, but also evolution, supercharged by **Chaon-Convolution events**. Because of **Epigenetic Triangle effects**, and the fact that selection acts on whole organisms, individual whole-genomic polymorphisms and continua in lineages are far more important than individual genetic polymorphisms within genomes. Many of the new genes imported by Convolution events, big and small, come fixed in their "best" (most useful) form, from previous Convolution cycles, contributing to the fitness of the new lineage or lineages created.

The **Convolution Cycle** consists of a series of processes starting with the **Convolution Event**, which reshuffles the nodal templates. Then follows a sequence, mediated by the Epigenetic Triangle, of Identification, Recognition, Acquisition, Positioning, Integration, and Utility of the new and/or reshuffled genetic material. [So-called "junk," or permanently silent genes, are Convolution leftovers, retained for awhile if they're not a burden. Many active genes are partially silent under various circumstances, or are parts of layered genetic control, cascade, and feedback mechanisms, and so act sporadically, intermittently, and/or differently under varying conditions involving other genes, somatic molecules, and/or environmental triggers]. This genetic intake, accommodation, rearranging, and testing is a primary purpose of the **Convolution Fluxus**, which follows the **Convolution Event**, and which typically involves a pattern of cross, backcross, and dilution (and, of course, much selective winnowing), observed results of which have been called **genetic introgression**. But it involves the entire **Epigenetic Triangle**, not just the genes.

Because there generally are continua of genetic possibilities, ranging from many forms of a gene to only one, including the creation at any time of new gene varieties through chaon-assisted mutation or importation by a Convolution event, the **Hardy-Weinberg equation** is not predictive over time beyond the interval between major mutation or Convolution events. This allows for extended Convolution to produce an **Interlocken** — a genetic combination, so tightly symbiotically bound to soma and environment through its particular combination of deep heritage and modern layered adaptations, that the organism produced becomes a relatively little-varying legacy line, a best-in-class for its time, the closest thing in nature to Plato's *Eidos* (in aircraft, the Piper Cub is a beloved and much-copied one, a "living fossil" that still flies somewhere every day). In other words, the "best" (and best-arranged) genes for that genome, vessel, environment, and time are fixed, and much of the variation has become internalized as Epigenetic Triangle-driven **Metamorphic Amplitude**, rather than genetic, polymorphism-driven variety. **Buck Moths are one of these lines; so are Humans**. There are also symbiotic Interlocken semispecies groups, which manifest as ecological community "types." Interlocken promote the appearance of order, stability, harmony, and beauty in Nature, qualities sorely tested now by rampant human activities and the consequent invasive and weedy release.

In order to persist, a new lineage, even one emerging from a legacy line, must be fit and nimble enough to **Escape the Fluxus**. This process is part of what's known as **Rolling Semispeciation**. It's how most semispecies, including our new Buck Moth, are born. A fluxus may also die out around one or more lineages (especially ones that are better adapted than the rest) and result in **Senescent Semispeciation**, but the circumstances for this are less common. This may have happened with the Buck Moths in and near Texas (*H. artemis, peigleri, & slosseri*).

Finally, the fluxus may keep expanding long enough to catch up with basal parts of the range of the new expanding semispecies, "swamping" it. This happened with our new semispecies, along with smaller secondary Convolution events within and near the original Great Lakes Fluxus, caused by later arrivals of other postglacial columns of recolonizing Buck Moths. **Secondary Fluxi** produced both northerly (through Michigan), and easterly (through Ohio and Pennsylvania to New Jersey) *shadow vectors*, with their own semispecies characteristics, and also produced and then diluted *Hemileuca lucina* (pp. 7, 26, 32). The pictures of that are in *Moths of Autumn*.

Humanity's rise is a classic case of Rolling Semispeciation through Convolution Cycles covering one large continent, Africa, over the past few million years, then spilling out across the rest of the Earth. In that short interval of geological time, a small arboreal ape turned into . . . us!

Long distances and large areas are required to make Convolution work. That is true for everything evolving in the concentric physical zones of the biosphere: air, water, and land, at all vessel scales, from single cells to whales and Sequoias. Because of this essential and existential demand of Convolution, the answer to the **Minimum Area Question** inspired by MacArthur & Wilson's 1967 *Theory of Island Biogeography* is *All of It*. *THE BIOSPHERE NEEDS THE ENTIRE EARTH TO FUNCTION*. The very nodal "Noah's Ark" approach of past conservation attempts, via limited preserves, is grounded on Mt. Ararat with a hole in its bottom. It was stranded there for good by global warming.

BEGINNING WITH HELPING TO USHER OUT THE PLEISTOCENE MEGAFAUNA, and in an exponentially intensifying onslaught over at least the past 50,000 years, we Humans have regressively and systematically, for our own self-centered purposes, unraveled, degraded, and dismantled the natural functions of our only Biosphere (*i.e.*, the "Older World"). And now, from our relative position of comfort and complacency, we are at the end stages of destroying it completely, through the ever-expanding "HIPPO" processes (*Habitat* destruction, releasing *Invasive* and *Introduced* semispecies, human *over-Population*, *Pollution*, and *Over-harvest* of many semispecies) that accompany our technology-enabled population explosion, turbocharged by rapidly accelerating global warming (RAGW), which resulted from burning up, in less than a few centuries, many millions of years'-worth of solar energy that was sequestered by past life. This has culminated in a grotesque orgy of frenzied production and overconsumption, without precedent or limit: buildings, vehicles, infrastructure, personal accouterments, junk food — essentially everything that can be imagined and made. Sacrificed on this dec-

adent altar to runaway consumerism are land, water, air, trees, fish, wildlife, and factory-farmed large animals.

Everywhere on the planet, ecosystems have been scrambled, and hollowed out by the losses of millions of semispecies. They have run down in all measurable functions, some easily noticed ones, besides semispecies diversity, being average body sizes, natural biomass productivity, and population sizes of everything but us and our companion animals, weeds, pests, and germs. We have done hideous cumulative damage to soils and waterbodies worldwide. Collapse is imminent (it has already started, in layered, cascading fashion, based on converging factors of vulnerability, rolling through fisheries, forests, coral reefs, insects, birds, large animals, crops, etc., down to a handful of semispecies, which will have to begin a long recovery). If RAGW is allowed to become *runaway global warming* (p. 40, *top*), life on earth will cease altogether, save for hardy, high-temperature remnants of the Cellosphere, which are the closest descendants of earliest life.

Because of *Genetic Overwriting* and *Chaonic Erasure*, Life is a Palimpsest. Through Convolution, accompanied by the vast winnowing power of *Enhanced Natural Selection*, driven by physics (the free chaonic forces of *Gravity* and *Time Inflation*), not just biology and chemistry, it rewrites itself over and over, reinventing and updating genes, genomes, life forms, and vessel features, innovations, and processes (including, especially, symbioses) to keep up with the continually changing times. Because there is no fossil record of almost all (at least 99.9999%) of the semispecies that have ever lived — the actual elements of evolution — we have fooled ourselves about what diversity there was, and what has actually been lost. And because of our nodal bias, we have underestimated the number of living semispecies still (some barely) holding on, by at least several orders of magnitude.

It would take a book-length treatment, and a lot of mock-continuum thinking, just to outline all the major implications of the *Chaon-Convolution Theory*. But the one that should focus the minds of everyone on the planet is this: Because of Chaons, Convolution, and Macrobiome and Microbiome Interdependence through Symbiosis (*the Master Life Process*), We're Not Getting Off Earth (and any other life forms out there are not getting off their planets, either). We are woven in. This is It. Our Only Home. For Good. In the immortal words of E. O. Wilson, "*HUMANS NEED A BIOSPHERE*." And not just any old biosphere: Our Biosphere! The essential one we evolved in, and remain utterly dependent on. *And We've Wrecked It!*

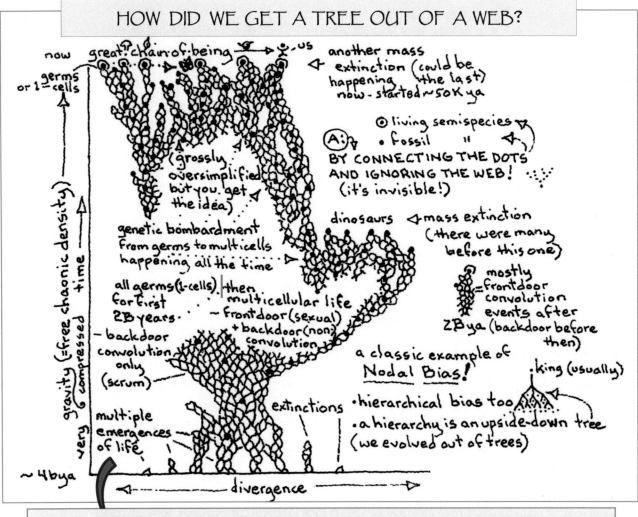

now or 1 = germs = cells

great chain of being → us

another mass extinction (could be the last) happening now - started ~50 Kya

⊙ living semispecies
• fossil "

A: BY CONNECTING THE DOTS AND IGNORING THE WEB! (it's invisible!)

(grossly oversimplified but you get the idea)

dinosaurs → mass extinction (there were many before this one)

mostly = frontdoor convolution events after 2Bya (backdoor before then)

genetic bombardment from germs to multicells happening all the time

all germs (1-cells) for first 2B years.

then multicellular life
~ frontdoor (sexual)
+ backdoor (non) convolution

- backdoor convolution only (scrum)

a classic example of Nodal Bias!

• king (usually)

• hierarchical bias too
• a hierarchy is an upside-down tree (we evolved out of trees)

multiple emergences of life

extinctions

gravity (= free chaonic density)   time   very compressed

~4bya

← divergence →

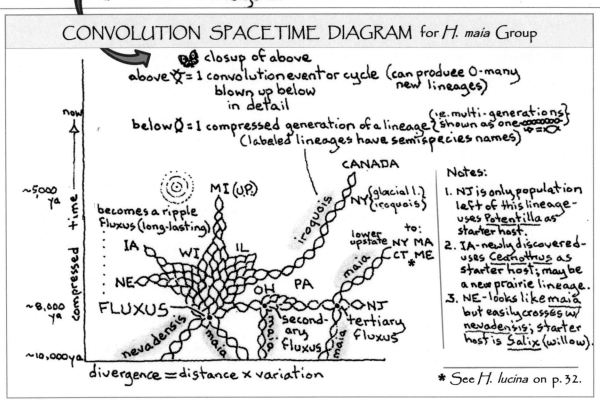

closup of above

above ⚥ = 1 convolution event or cycle (can produce 0-many new lineages) blown up below in detail

below ♀ = 1 compressed generation of a lineage (ie. multi-generations shown as one) (labeled lineages have semispecies names)

now

~5,000 ya

becomes a ripple Fluxus (long-lasting)

CANADA

MI (U.P.)

NY (glacial l.) (iroquois)

iroquois

IA   WI   IL

lower upstate to: NY MA CT ME *

maia

NE

PA

OH

FLUXUS

~8,000 ya

nevadensis   maia

second-ary fluxus

maia   NJ tertiary fluxus

~10,000ya

time   compressed

divergence = distance × variation

Notes:
1. NJ is only population left of this lineage - uses Potentilla as starter host.
2. IA - newly discovered - uses Ceanothus as starter host; may be a new prairie lineage.
3. NE - looks like maia but easily crosses w/ nevadensis; starter host is Salix (willow).

* See *H. lucina* on p. 32.

The good news is, it's the eleventh hour, but it's not too late. Convolution is so powerful a force that it can do wonderful things very fast in geologic time, if we bring it back to life. We Need to Allow our Biosphere to Reclaim Itself. That means We Need to Give It Back Its Space. And that means **UNDER-REPRODUCE** by one-half for *three* (severely over-populated areas will take *five*) generations, in order to Rapidly Diminish, and Get Out of most of it. This will require all of Humanity to curb its animalistic instincts, emotions, appetites, desires, and drives in a coordinated, disciplined, and sustained effort, the likes of which has never been attempted before. It will be supremely difficult to do and sustain, because it goes against parts of our deepest layered-in evolutionary survival strategies that may have worked in the past, but are killing us and the rest of life now. Shared sacrifice and suffering will be part of this, too, at least initially. Most cities, suburbs, towns, factories, farms, ranches, dams, mines, and roads will have to go, in our quest to **UNDEVELOP THE EARTH** and return it to wilderness. (Coastal areas are already doomed by the momentum of sea level rise.) An enormous portion of our wealth must be poured into perma-nently altering the ways we live, and our re-lationships with the environment that has nurtured us, until we ate and used it up. We will also have to abandon many of the fundamental technological arts that made us human (especially the ancient and modern Promethean ones, as well as the fatally flawed digital diversion we recently foisted upon ourselves; it can and will kill us, all by itself), and invent and deploy new ones in their places.

Here are **Ten More Commandments** (the old ones still apply, although needing some updated language). They came from the mire, not the mountaintop! No matter, they came from God. The one, constantly growing, compound-compositing God who cares for all of us, and whose inklings have appeared in many manifestations over our short history. The God who is invested in us be-cause **Hre** (that's a gender-inclusive pronoun) created the Universe that created us — and God expects us back. It's the God we can reach and return to by way of the love and rescue we practice here on Earth, for one another, and for Creation. The alternative is spiritual waste: eternal loneliness, far in the outer void, the true Hell. With not a Chaon for company.

Follow these, and we'll have a future. Here and in there. *Buckies Forever!*

℘℘

TEN MORE COMMANDMENTS FOR HUMANITY AND OUR ONLY HOME

℘℘

XI
DIMINISH: HUMANELY REDUCE YOUR NUMBERS TO UNDER 1 BILLION WORLDWIDE AND KEEP THEM THERE.

XII
RETURN 7/8 OF THE LAND, AND ALL WATERY REALMS, TO NATURE, UNTRAMMELED BY HUMANITY.

XIII
MAKE WOMEN EQUAL TO MEN, AND ALL PEOPLE EQUAL TO ONE ANOTHER.

XIV
CREATE AN EQUITABLE ECONOMY NOT DEPENDENT ON PHYSICAL GROWTH.

XV
NEITHER MAKE NOR UNEARTH ANY CHEMICAL OR THING THAT HARMS THE BIOSPHERE.

XVI
CREATE NONLETHAL WEAPONS FOR SECURITY AND POLICING, AND DESTROY THE REST.

XVII
ABOLISH WARFARE OF ANY KIND.

XVIII
BURN NOTHING BUT HYDROGEN.

XIX
RESPECT AND CHERISH ALL FELLOW BEINGS.

XX
REMEMBER THAT I AM WAITING FOR YOU TO REJOIN ME AND YOUR LOVED ONES.

Epilogue

New Atlantis, 2258

JAMIE CAMERON ANGELUS stared with her long glass through the fog enveloping the bridge of WAVERUNNER ARC VII. It was daybreak, and the swirling mists had just started to rise. Gulls had been shadowing the ship for more than a day now. Sure enough, she made out a golden light off the port bow. There was only one light in all the world like that. She was home.

As the wisps dissipated above rolling chop, more, paler lights showed to starboard, one at a time, in a string fading back into grey murk. She adjusted course slightly to pass between the first one and the line. Afterward, she raised a finger and began speaking to a tiny bedraggled creature hanging from it. "I still got it, Scrappy. Blind-navigated us all the way across, and didn't miss by even a mile." The little black animal raised one of his legs and seemed to wave it at her.

She put the ship on auto and went below to make some coffee. "Time for bed," she said to her finger, then gently let her night companion crawl off into a moss-lined container, which went into a special cooler. "Thanks for keeping me company."

When she came back, the ship had maneuvered itself between the two nearest lights. There was enough pre-dawn glow to make out the familiar figure holding aloft the gold one, and the equally recognizable top of a sky-scraper bearing the other on its spire. The former crowned a low-slung mound covered in huge stone pieces; the other simply jutted from a wide expanse of sea.

Jamie retook the helm as WAVERUNNER passed the Fresh Kills Island sound buoy. She centered the thousand-foot septuagent sesquimaran squarely over the midline of New York Trench, and trimmed the thin, titanium, helical, junk-rigged sails on her five telescoping masts. When the ship passed well port of the Empire Light and the clustered Lost Apple Salvage barges over Manhattan Reef, she breathed a sigh of relief. She was carrying precious cargo. Any delay, never mind the calamity of a gouging or grounding, could spell its loss.

As she entered the Hudson Fjord proper, Jamie went to the starboard bridge wing and looked out. A blood red ball had just broken horizon. It glittered astern on the silvery, hollow window frames of the famous building-turned-lighthouse. Twin irregular, diverging lines of small islets extended eastward, vanishing in sunrise haze. Some of the southern line had lights, now dimming as day took hold. The largest, High Hill Island, though far off, had the brightest light. North of the islets, a distant shoreline of waves breaking on low cliffs diminished to the northeast horizon.

Back at the helm, the expanse of the fjord widened northward. Small waves broke to the immediate west over Palisades Shoals; beyond, a calm sea spread almost as far as she could see, to the First Watchung Ridge. Jamie decided this was a good time to take a nap. She'd been up almost twenty-four hours.

While Jamie slept, WAVERUNNER passed effortlessly into and through Tappan Sea, a vast inland lake and the widest part of the fjord. In parts of the passage, no familiar land could be spied close at hand. No matter; the ship always knew precisely where it was. Jamie was the backup, unless she was sailing the old-fashioned way.

The ship woke Jamie when it was time to thread the shallows between Storm King Island and the Schunemunk Peninsula. Beyond that, the fjord widened again westward, coming up hard against the low Shawangunk Ridge and beyond it, the still-formidable, cobalt-blue Great Wall of Manitou. The sun was now headed toward zenith on a beautiful late fall day.

The brightly-colored, revegetated burn scars on the slopes of the Wall reminded Jamie it was time to inspect the cargo. They'd be landing soon. She put back on auto and went to the rear stairway. Before her, spaced between the masts and resting on the seven lightweight carbon composite hulls, were four 200-foot-wide geodesic biome domes. Inside, she could see each meticulously grown and assembled ecosystem and its little robot attendant buzzing about: *taiga, tundra, bog,* and *fen*. To visit each was to step three centuries back in time.

These ecosystems did not exist anymore. At least not on the surface of the Earth. They had been reconstituted by ARC and its many collaborators from seed stock collected by prescient people and stored under the mountains of Norway. As the rapid warming progressed and began to accelerate, more and more people got involved in biosphere salvage, collecting seeds, spores, animals, insects, frogs, sea life — even microbes — from dying natural areas worldwide, and figuring out ingenious ways to keep them alive in some form for this moment. In the end, through all the horrors that had ensued in those benighted years, millions of people had risen up to literally save Creation.

And now it was time to stand and deliver. The politics had stabilized. Nuclear weapons and power were a thing of the past. Conventional weapons were in rapid decline. The many smaller nations created by Boundary Redraw Day had settled into the mere administrative states they were meant to be. The United States was now basically forty-nine very friendly states, some mere remnants of their former selves. A "slow economy" had taken hold. The human population was in controlled free-fall: under *three* billion and dropping toward a stable *one*. For nearly two hundred of those prior years, the losses had been uncontrolled.

It had been an incredibly close call. Runaway global warming had barely been averted. The seas had risen almost three hundred feet, and were beginning to vaporize into the atmosphere.

The tide had turned within a few years of the sealing of the last oil and gas wells, and burial of the last coal mines. It began snowing again all winter, starting at the poles. The ice sheets began to re-form, then grew, slowly and tentatively, a little farther each winter.

And this past year, for the first time in what seemed an eternity, the seas halted their rise, and in a few places dropped an imperceptible amount.

So the call was made: WAVERUNNER would make the maiden voyage, from the Old World to the New. Just as in the past. Only this time to return, not take. And Jamie, proud descendant of an unbroken line of distinguished mariners going back seven generations, would helm her colossal invention.

Jamie finished her tour of the domes in her favorite. *Newy* came out to greet her, as had his siblings *Huey, Dooey,* and *Louie.* He beeped and led Jamie through the moss, sedge, and waving cottongrass to a tiny pond opening: the "eye" of the fen. There, on an open Buckbean seed-pod, was a gift. Not for her, but for her little pet back in the cabin.

Through the honeycomb glass Jamie watched the Great Wall recede, and the lower Helderberg escarpment came into view. Time to get back to the bridge. Port call was just around the corner. And cargo handoff onto arctic-passage vessels to final destinations.

Jamie took the con and altered course to port, trimming sail as she did, sending WAVERUNNER westward over Normanskill Channel and bypassing Fort Orange Reef. Ahead, in the distance, lay the serried, undulating dunes of Pine Bush Landing, covered with Southern Hard Pines rising over dense low thickets of Blackjack, Post, and Turkey Oak. And for the first time in many years, she set eyes on her birthplace, rising on the limestone cliffs behind the pine barrens. Its high towers shone in the streaming sunlight, new and old standing together. Tears came to her eyes.

Then a mighty roar rose up, a sound like none she'd ever heard: majestic, deep, and rolling. And Jamie realized with a start that an enormous dark mass along the landing was not an afternoon dune-shadow. It was *people*, an immense crowd, filling the garlanded quay and spilling over into the wilderness beyond. The entire city, it seemed, had turned out to meet WAVERUNNER. Her heart caught in her throat.

She had presence of mind to furl sail and set the docking sequence. Then she remembered what she had in her pocket and bolted down to her cabin.

Scrappy emerged from his nest, and without hesitation began to climb up the proffered newcomer's tail. Love at first sight — and smell. The little guy, who'd been deprived of properly expanding his wings by a stormy night at sea, would not be denied his destiny.

Jamie threw on her formal jacket and captain's cap and raced back to the bridge with one last act to perform, before being engulfed by her welcome.

She snapped a tiny drive into her custom sound system. From dozens of speakers hidden in the masts came a wall of music. It was from the American Classical Period. The crowd slowly went silent as the familiar words and melodies rolled over the people. Then, as one, they sang along, their voices ringing back from the city. *Love Is All You Need. So This Is Christmas. War Is Over. Imagine.*

And then a single, black-white-and-red moth flew in an undulating, up-and-down pattern over the landing.

৪০৫৪

Jamie and "Scrappy"

Addenda

THE LAST STAR: MAIA IN MENOPAUSE, AT THE END OF TIME, WITH HER CHILDREN ARCADIA AND HERMES

[*Also see frontispiece & p. 15.*]

Why *Maia*? It will remain an eternal mystery why DRU DRURY picked the appellation *maia* to grace forever the beautiful moth[1] that arrived by packet from old New York on the cusp of revolution. As the taxonomists of antiquity all had classical training, perhaps it was the mystery that shrouds Maia herself, to this day, as a major member of the pantheon about whom little original *story* survives. In this chaonic deficit of posterity (as STEPHEN FRY has wryly noted and slyly exploited[2]) lies opportunity. The Greeks have generously bestowed their gods upon Creation to live on in the consciousness of the world, as *prima manifesta* of the many faces of the one composite God. And so we have grabbed the opportunity — nay, coincidence — to steal once more the ideal form of Maia, giving her new life and meaning, and setting her once again among her heavenly sisters, the stars, to tell the true story of life and the mortal cosmos that birthed us. In this appointed task, we have been blessed by Hermes himself, son of the union of Maia and Zeus, and the god of, among many causes, those of thieves.

NOTES: 1) *Hemileuca maia* (Drury), 1773, the type species of that genus. 2) *Mythos, The Greek Myths Reimagined*,
Chronicle Books, San Francisco, California, 2017, 351 pp.

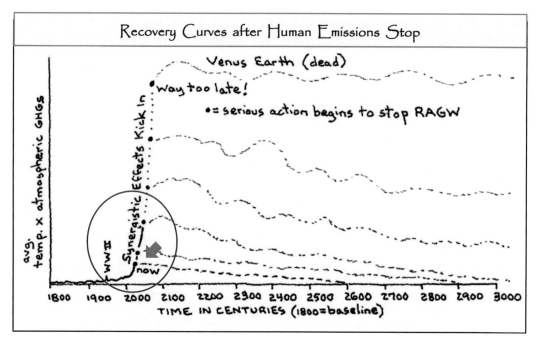

Recovery Curves after Human Emissions Stop

Venus Earth (dead)

Way too late!

• = serious action begins to stop RAGW

avg. temp. x atmospheric GHGs

Synergistic Effects Kick In

WWII

now

TIME IN CENTURIES (1800=baseline)
1800 1900 2000 2100 2200 2300 2400 2500 2600 2700 2800 2900 3000

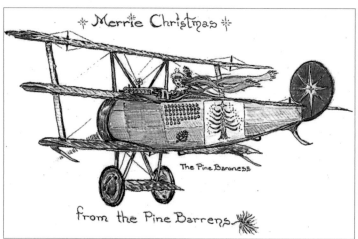

✳ Merrie Christmas ✳

The Pine Baroness

from the Pine Barrens. ✺

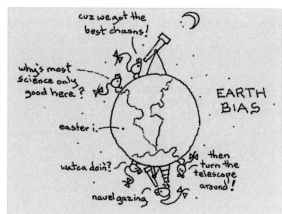

cuz we got the best chaons!

why's most science only good here?

EARTH BIAS

easter i.

watca doin?

navel gazing

then turn the telescope around!

Ʌ **"THE PINE BARONESS"** was a mascot for the "War of the Woods" (1989-1993), the Long Island Pine Barrens Society's campaign to save 100,000 acres of this rapidly vanishing habitat. It involved the largest land use lawsuit ever brought, followed by decades of ultimately successful negotiations, politicking, and additional legal action, involving many public and private entities, to create the preserve. The Baroness' Fokker Triplane was probably the most iconic aircraft of WWI. It was slow but incredibly maneuverable, giving it the advantage in one-on-one combat. It was eventually outclassed by faster opponents, using coordinated tactics that would be fully developed in the next world war. [*See pp. 18-19.*]

Bog Buck Moth flight sequence on pp. 40-42

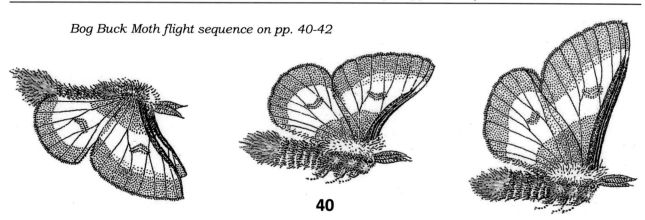

The type locality in the 1970s

Recommended citation for this publication:

CRYAN, JOHN F., & ROBERT DIRIG. 2020. *Moths of the Past: Eastern North American Buck Moths (Hemileuca, Saturniidae), with Notes on Their Origin, Evolution, and Biogeography. Pine Bush Historic Preservation Project Occasional Publication No. 2,* 44 pp., covers, 80 illustrations.

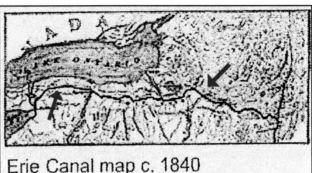

Erie Canal map c. 1840

⋏ Construction of the **Erie Canal** (*arrows*) presaged the end of most of the fens of ***Glacial Lake Iroquois,*** and the mires of the Midwest. (*Image courtesy of Wikipedia Commons.*)

During postglacial migration, old and new lineages of the Buck Moth (*maia, left*) colonized many new habitats, and became so widespread and abundant that the Red-spotted Purple quickly parlayed a simple polymorphism into the White Admiral *metamorphic variety* as a Buck Moth mimic (*center & right*), following the distasteful and poisonous moth model as it moved north.

A Senryu

we are as wind in
the grass — triumphant — then the
grass grows still again

~ *anonymous edo poet*

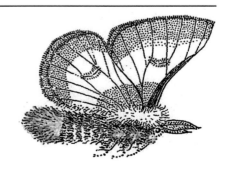

IN THE LAB OF DR. FRANCLEMONT, 1977

And so John began a senior research project to solve the identity of the mystery moth.

J. G. FRANCLEMONT was the last of the "American Oxbridge" dons. Despite his name, he was an anglophile through and through, and keeper of the flame of his formidable mentor, W. T. M. ("Pinkie" to his friends) FORBES. He occupied Forbes' old upstairs warren on the west side of venerable Comstock Hall, home of the "pure" part of the Cornell Entomology Department.

Dr. Franclemont was proper, unfailingly polite, courteous, and generous to a fault, especially with his time and knowledge. He could be found day, night, and weekends in his office or lab. His moth collection was renowned for the impeccable perfection of the specimens in long series and the Professor's exacting rearing and mounting techniques. No one else had the patience, discipline, and old-fashioned, hand-eye talent to produce such bespoke Victorian artifacts. A conversation with him often involved consulting his beautiful and extensive personal collection, which he kept in overflowing rows of cabinets under his watchful eye, separate from the "disheveled" university one. It also, if one were lucky, could result in being handed a first edition jewel from his equally famous lepidopterological library, which contained volumes hundreds of years old, going all the way back to Linnaeus and beyond.

He was also an outstanding dissectionist. Beginning graduate students were sent from other top colleges to learn how to remove, clear, prepare, and mount insect genitalia on glass slides. That was the first set of skills John learned under his tutelage, spending long winter hours with an old Bausch and Lomb microscope that Vladimir Nabokov had used in the 1940s in a small side lab, filled with the smell of distilled Canadian balsam and oil of cloves, with the professor's little white yorkie, Choe, sleeping underfoot.

Dr. Franclemont's fundamental taxonomic method was little changed since the great eruption of the real Victorian era produced by Charles Darwin. It was called "lock and key," based on the notion that if two otherwise seemingly close organisms couldn't mate for reasons of mechanical incompatibility, they must belong to separate species. This worked wonderfully in the Professor's chosen family, the owlet moths, a hyper-diverse array of mostly common porch light visitors whose eyes glowed gold or even red in the shadows. The males had baroque ornamented armatures for genitalia that looked like they belonged on miniature dragons. And the females had corresponding receptacles to accommodate that particular configuration. There were thousands of variations of "locks" and "keys."

John was disappointed to find out that this was not the case for Buck Moths. The males' parts were basically an almost featureless capsule, and the females' nothing more than a tube leading to a sack. They were more or less indistinguishable among the three named species and their geographic variations, and they were more or less the same size. The new moths' looked basically the same as all the old ones. It was pretty clear that, mechanically speaking, any of them could mate with any of the others.

Dr. Franclemont consoled John by saying it was a good project anyway, that science work mostly resulted in failures, and that meant we had to look elsewhere for answers. He left as his parting words a classic Franclemontian aphorism, pithy and piquant:

"We may name them, or not, but they know who they are."

༄༅

◀ Mated pair of *Hemileuca iroquois* (*male on left*), 1970s. ▲ A male *iroquois* in aposematic display, landed on a twig, 24 Sept. 2017, Lake Ontario East Shore.

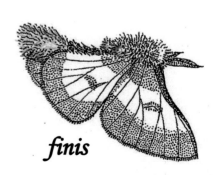

finis

Authors & Illustrators

John F. Cryan and Robert Dirig met at Cornell University in 1973, where both studied entomology, and began to collaborate on field projects with Lepidoptera, plants, and lichens. They researched Buck Moths, Karner Blues, Silvery Blues, and pine barrens habitats throughout the Northeast, and were involved in early conservation efforts of the Xerces Society. John co-founded the Long Island Pine Barrens Society, which helped preserve this endangered ecosystem on L. I. Both were instrumental in helping to safeguard essential tracts of the Albany Pine Bush in N.Y., and were also active in the Teen International Entomology Group in the 1960s-1970s.

John is retired from a lifelong career with the N.Y.S. Dept. of Environmental Conservation, and Bob has been a botanical and mycological curator and natural history educator at Cornell University for many years.

❧☙

John Cryan (left), watching for Bog Buck Moths at a fen in Oswego Co., N.Y., on 21 Sept. 1984; and Robert Dirig (right), holding John's box of Buck Moth pupae and emerging adults, at a Great Lakes Buck Moth Fluxus site in Monroe Co., Michigan, in Sept. 1987.

Afterword & Acknowledgements

The extraordinary happenstance and opportunity of John's very thorough study of the North American *Hemileuca maia* group, over half a century, have provided enough information and perspective to yield this interpretation of their evolution and biogeography — and ultimately to produce his *Chaon-Convolution Theory*, which provides a deep context for many aspects of Earthly life and its processes.

We did not intend to disrupt science, but once gravity is flipped from *pull* to *push*, everything reverses. This applies especially to philosophy.

What, exactly, has changed? Is it art, of which science is a part? Is it Humanity itself? And do we have a future, without modern science and its technological spawn?

We would answer *yes, yes*, and *yes*.

We have defused art, and science, as religions, and now are free to begin once more. This last time, we need to get it right, as we are almost out of time, which can never be hoarded.

Humanism, the receptacle for all this art and science (in its manifold individualistic, secular, social, liberal, conservative, and additional forms) is our DOA-last-gasp at religion.

Like all other religions before this, it developed a fatal flaw: *hierarchy*, with a clergy, separate and above people, which protects only its own interests, using secret rituals, traditions, and languages. This has warped our view of the world, as our collective knowledge comes out of those things, but we do not understand it. The hierarchs want to keep it that way. Their latest foray, the Internet, promises to do just that, in the guise of doing the opposite. Unless it is abandoned in its current forms, the destruction of our planet as a place for Humans and much other Earthly life is assured.

Beyond that, we must eliminate hierarchy itself. It is the worst of many bad nodal concepts, because it is completely artificial. Nature works through continuously operating webs of relationships. Each strand, or continuum, is vital. And equitability rules Nature. We all stand equal before God. And God is indeed very much alive, and growing.

So what is to become of science, shorn of the hegemony of math, in a self-erasing mirage of a Universe? There is only one answer: Fallen from its position as techno-hierarch of the arts, science must rejoin them. People cannot gain true perspective from inside a subject — perspective always comes from a distance (and sometimes from the added distance of time). Artist-scientists must reunite with their artistic sistren and brethren. And since all Humans are artists, we are all scientists, too. This especially includes journalists, historians, and other observers — those intrepid seekers of nodal truths of the moment and the past, from which we can then build larger, more subtle, and continuous certainties. It is also time to acknowledge and resurrect the vital role of amateurs in science.

We hope that this small effort of ours, despite any faults or omissions, will convey some essential truths. As authors, we acknowledge our debts to the entirety of Humanity and to Nature, as experienced through two individual lifetimes. Any attempt to list all persons and sources to credit would be incomplete; but we need to mention a few individuals whose presence and work have helped us turn a tiny moth paper into an outsized philosophical treatise. In addition to a few, who were already named in our text, we thank the following:

PETER ROSENBAUM and his colleagues, who preserved the few remaining fens on the east side of Glacial Lake Iroquois; SANDRA BONANNO and her collaborators, for steadfastly witnessing and documenting the decline and loss of several of the lake bed populations of *Hemileuca iroquois* in the face of climate change; KAREN SIME, who gave generously of her time and knowledge to support the difficult gestation and writing of this document, and facilitated our 2017 return to the original site we discovered, exactly forty years later, to re-experience and enjoy the mystique of this lovely small saturniid in its sublime habitat, throughout a glorious autumn afternoon. We also acknowledge the help of JASON DOMBROSKIE, who agreed to repatriate our voucher specimens; TORBEN RUSSO and DON RITTNER, who provided critical support for producing this work; CAROLYN KLASS, who offered a helpful review; and DANIEL RUBINOFF, who gave us encouragement and help that was above price. Finally, we dedicate this paper to the late JOHN G. FRANCLEMONT, our early Teacher, Mentor, and Friend, in appreciation for his guidance and inspiration.

ଙଊଓ

➤ BACK COVER: *JUNE NOCTURNE IN THE FEN*: Borrowing from M. C. ESCHER's famous *Three Worlds* lithograph (1955), this recent scene encompasses *four* worlds in one: **Bog Buck Moth larvae** feed on a **Buckbean** leaf that emerges from a Sphagnum carpet, alongside its intricate flower-spike with young fruits at the base; a **Water Strider** (Gerridae) dimples the surface film; a submerged, predaceous **Giant Water Bug** (Belostomatidae) stalks the Strider, while a tattered **forewing of the Moth** that froze in ice over the winter drifts by; and the starry sky and full moon frame a mournfully cherishing **Manitou** in the water-mirror, manifesting a proto-Iroquoian presence. Also in the foreground, an **empty egg ring** and **pupal shell** of the Moth finish the shape-shifting guises of a dramatic metamorphosis.

ଙଊଓ

Tailpiece

To the Edge of the World

෭෨